The Institute of Biology's
Studies in Biology no. 66

Lichens as Pollution Monitors

David L. Hawksworth
Ph.D., F.L.S.
Mycologist, Commonwealth Mycological Institute, Kew

Francis Rose
Ph.D., F.L.S.
Reader in Biogeography,
King's College, University of London

Edward Arnold

© David L. Hawksworth and Francis Rose 1976

First published 1976
by Edward Arnold (Publishers) Limited,
25 Hill Street, London, W1X 8LL

Boards edition ISBN: 0 7131 2554 3
Paper edition ISBN: 0 7131 2555 1

All Rights Reserved. No part of this publication
may be reproduced, stored in a retrieval system,
or transmitted, in any form or by any means, electronic,
mechanical, photocopying, recording or otherwise, without
the prior permission of Edward Arnold (Publishers) Limited.

Printed in Great Britain by
The Camelot Press Ltd, Southampton.

General Preface to the Series

It is no longer possible for one textbook to cover the whole field of Biology and to remain sufficiently up to date. At the same time teachers and students at school, college or university need to keep abreast of recent trends and know where significant developments are taking place.

To meet the need for this progressive approach the Institute of Biology has for some years sponsored this series of booklets dealing with subjects specially selected by a panel of editors. The enthusiastic acceptance of the series by teachers and students at school, college and university shows the usefulness of the books in providing a clear and up-to-date coverage of topics, particularly in areas of research and changing views.

Among features of the series are the attention given to methods, the inclusion of a selected list of books for further reading and, wherever possible, suggestions for practical work.

Readers' comments will be welcomed by the authors or the Education Officer of the Institute.

1975

The Institute of Biology,
41 Queens Gate,
London, SW7 5HU

Preface

The impact of man on both our physical environment and our flora and fauna is currently receiving more attention than ever before. A great deal of research into these topics has now taken place but the bulk of this remains in publications not readily accessible to students and teachers. An introduction to biological aspects of pollution has been included in an earlier booklet of this series (K. Mellanby, *The Biology of Pollution*, Studies in Biology No. 38, 1972). The present number aims to provide a synopsis of the effects of pollutants on one particular group of organisms, namely the lichens. Lichens are an enigmatic group of plants in many ways and frequently receive only scant attention in school and university courses. They can, however, be used to illustrate the drastic effects of man on a group of organisms and also the ways in which these effects can be of service to man in ascertaining the levels and patterns of pollutants. In addition to briefly reviewing theoretical aspects of the topic, this booklet emphasizes phenomena which can be seen easily in the field and describes some projects suitable for use in schools, colleges and degree courses.

Kew and London, 1975

D. L. H.
F.R.

Contents

1. **What are Lichens?** — 1
   1.1 Introduction 1.2 Fungal partners 1.3 Algal partners 1.4 Structure 1.5 Reproduction 1.6 Growth 1.7 Physiology and synthesis 1.8 Ecology and distribution

2. **Sulphur Dioxide** — 9
   2.1 Sulphur accumulation 2.2 Effect on photosynthesis and respiration 2.3 Effect on vitality 2.4 Effect on substrate preference 2.5 Mechanisms of resistance

3. **Other Pollutants** — 15
   3.1 Smoke 3.2 Fluorides 3.3 Car fumes, hydrocarbons and dust 3.4 Metals 3.5 Radionuclides and radiation 3.6 Agricultural chemicals 3.7 Freshwater and maritime pollution

4. **Other Factors affecting Lichen Distribution** — 21
   4.1 Introduction 4.2 Woodland management 4.3 Heathland management 4.4 Man-made substrates 4.5 Drought 4.6 Topography 4.7 Climate 4.8 Public pressure

5. **Mapping Air Pollution Patterns** — 26
   5.1 Introduction 5.2 Species mapping 5.3 Zone mapping 5.4 Other methods 5.5 Indicator species and standardization 5.6 Correlation with pollutant levels 5.7 Interpretation and value

6. **Impact of Sulphur Dioxide on the British Lichen Flora** — 37
   6.1 Historical basis 6.2 Declining species 6.3 Increasing species 6.4 Species numbers 6.5 Other effects

7. **Other Considerations** — 43
   7.1 Air pollution trends 7.2 Effects on other plants 7.3 Effects on man and his materials 7.4 Effects on other organisms 7.5 Conservation

A1. **Appendix: Lichen Identification** — 46
   A1.1 Collection and preservation A1.2 Conservation A1.3 Naming A1.4 Some species used in air pollution surveys: leprose and crustose species, squamulose species, foliose species, fruticose species

A2. **Appendix: Practical Exercises** — 56
   A2.1 Introduction A2.2 Species numbers A2.3 Distribution of growth forms A2.4 ACE zone scale A2.5 Transplants A2.6 Detailed species mapping and listing

Further Reading — 60

1 What are Lichens?

1.1 Introduction

Lichens (pronounced *lie'ken*, from the Greek Λειχήν=tree moss) are not a single group of plants comparable to, for example, mosses and liverworts (Bryophyta) or ferns (Pteridophyta), but are composed of two quite different organisms, a fungus and an alga. They are essentially fungi united only in having a common method of nutrition, symbiosis with algae. The resultant compound organisms behave as if they were single biological units and differ in so many ways from their free-living relatives that for convenience they are often studied as if they were a single natural group.

Lichens are an outstandingly successful group of symbiotic organisms exploiting a wide range of habitats throughout the world. There are certainly in excess of 18 000 species belonging to about 500 genera in the world; 1368 species are known in the British Isles.

1.2 Fungal partners

The fungal partners (mycobionts) of lichens are mainly Ascomycetes but a few Basidiomycetes (two in Britain), some Fungi Imperfecti (tropics) and possibly one Phycomycete (central Europe) also form lichens. Some concept of the success of the lichen-forming Ascomycetes may be obtained from the realization that over half of all known Ascomycetes in the world are lichens. Lichen-forming and 'free-living' (non-lichenized, i.e. saprophytic and plant parasitic) fungi have been studied and classified independently for over two centuries and relationships between them are poorly understood. Evidently, lichen mycobionts have been derived from a number of distinct groups of free-living Ascomycetes (some perhaps long since extinct). A few genera include both lichenized and non-lichenized species (e.g. *Buellia*), but the latter may have been derived secondarily from lichens by the loss of the need to depend on an alga for their nutrition. Most lichens do not now have close relatives amongst the non-lichenized Ascomycetes; while many have disc-like fruits recalling those of the Discomycetes ('cup fungi') their detailed structure is generally very different.

Mycobionts can be isolated and grown in the laboratory free of their algal partners but they then usually only form amorphous masses of sterile mycelium quite different from the characteristic shapes of the intact lichens from which they were derived. Isolated mycobionts are hardly ever found in nature and it is clear that they need an appropriate alga to survive under natural conditions.

Strictly speaking, the names of lichens refer only to their fungal partners so that the associated algae can bear their own Latin names. The mycobiont usually forms the bulk of the compound structure and is the sole method of sexual reproduction. The algae, however, have an important rôle to play not only in nutrition but also in the determination of thallus form. A few cases are known where a single mycobiont can form quite different structures with different genera of algae.

1.3 Algal partners

In contrast to the fungal partners, the algal partners (*phycobionts*) mostly belong to genera well known in the free-living as well as in the lichenized state. Green algae (Chlorophyceae) predominate, but blue-green algae (Cyanophyceae) are the main algal components in some genera. A few lichens consistently have both a green and a blue-green alga (the latter often confined to secondary vegetative structures termed *cephalodia*), but the majority have one or the other.

The green algae are most frequently unicellular forms (e.g. *Trebouxia*), but some lichens have filamentous algae belonging to genera such as *Trentepohlia*. Free-living *Trentepohlia* species are often seen as orange-yellow fluffy patches on damp shaded walls, while *Trebouxia*, in contrast, is rarely encountered outside lichen propagules in nature. The commonest blue-green alga in lichens is *Nostoc*, free-living species of which form jelly-like masses on limestone rocks and heaths after rain. The number of algal species known to occur in lichens is extremely small compared with the fungal species involved. While fungal partners will only form lichens with one, or rarely two, particular genera of algae, there are indications that they are less specific as to the species or strains of algae within those genera.

1.4 Structure

The fungal and algal partners together constitute the lichen *thallus* which may assume a variety of forms depending on the fungal and algal partners involved. The main types of thalli (Fig. 1–1a) are *leprose* (powdery mass of algal cells and fungal hyphae with little or no organized structure), *crustose* (crust-like, with the algae lying below a distinct layer of fungal tissue termed the *cortex*), *squamulose* (crustose but with the margins becoming raised from the substrate; *placodioid* if furrowed and lobe-like), *foliose* (leaf-like, with a distinct upper and lower cortex, often attached to the substrate by means of hair-like rhizinae and easily peeled off it), and *fruticose* (erect shrubby or beard-like pendent species attached to the substrate only at their bases, usually circular in section). These various types are not indicative of relationships; while most lichen genera include only a single type, one family can include genera with quite different types.

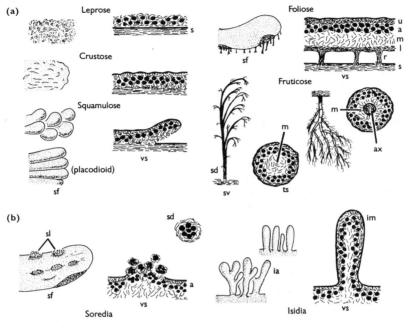

Fig. 1–1 (a) Growth forms and thallus structure; (b) vegetative reproduction. a = algal layer; ax = axis; ia = groups of isidia (×40); im = isidium (×75); l = lower cortex; m = medulla; r = rhizinae; s = substrate; sd = soredium (×150); sf = surface view (×1); sl = soralia; sv = side view (×½); ts = transverse section (×20); u = upper cortex; vs = vertical section; ● = algal cells. (Diagrammatic.)

The detailed structure within the main types of thallus form shows considerable variation in different genera. In Fig. 1–1a the thallus in all but the leprose species will be seen to be layered (*stratified*), but a few genera belonging to other growth forms (e.g. foliose) have thalli with algae and fungal hyphae intermixed throughout the thallus (*unstratified*; e.g. *Collema*). The layers of the thallus are most easily demonstrated in foliose species: a dense, often pseudoparenchymatous, layer of fungal hyphae forms the upper surface (*upper cortex*) which is frequently pigmented; the algae occur in a rather irregular layer internal to this where the fungal hyphae are less dense (*algal layer*); the region below the algae generally consists only of a loose net of fungal hyphae (*medulla*); and the lower surface (*lower cortex*) may or may not be similar in structure to

4 REPRODUCTION

the upper cortex but is often thinner, differently pigmented or not pigmented at all, and in some cases has attachment organs (e.g. rhizinae) arising from it.

1.5 Reproduction

Different species of lichens may have one or more of several types of reproductive structures. Sexual reproduction (Fig. 1–2) is by means of *ascospores* violently ejected from sack-like structures (*asci*), each of which usually includes eight ascospores (Fig. 1–2h). The asci may be arranged in a variety of ways in cup-shaped or disc-like (*apothecia*), elongate or script-like (*lirellae*), or almost globose (*perithecia*) structures, the general term for which is *ascocarps*. Ascocarps may be immersed in the thallus, originate on

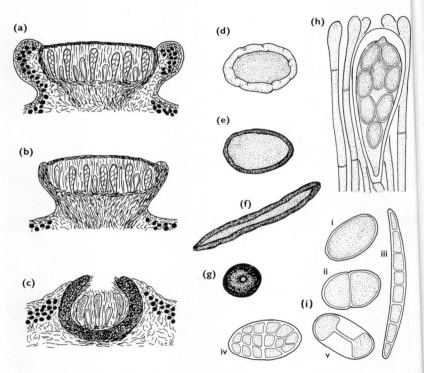

Fig. 1–2 Ascocarp types and structure. (**a**) Lecanorine apothecium (vs); (**b**) lecideine apothecium (vs); (**c**) perithecium (vs); (**d**) lecanorine apothecium (sv); (**e**) lecideine apothecium (sv); (**f**) lirella (sv; vs as B); (**g**) perithecium (sv); (**h**) ascus and paraphyses; (**i**) ascospores (i, simple; ii, 1-septate; iii, multiseptate; iv, muriform; v, polarilocular). sv=surface view; vs=vertical section. (Diagrammatic.)

its surface, or be stalked. The detailed structure and method of development of ascocarps varies very considerably and it is by these variations that lichens are primarily classified into orders, families and genera.

Ascocarps are not known in all species and many others form them only rarely. Such species rely on vegetative (asexual) methods for their propagation. Unlike ascospores, which on germination give rise to the fungal partner alone (and so must find a suitable alga in order to form a thallus), most asexual methods of reproduction disperse elements of both algal and fungal partners in a single propagule (and so can give rise directly to a thallus). The abundance of lichens reproducing only by asexual methods testifies to their efficiency.

The simplest method of asexual reproduction is by thallus *fragmentation*; small pieces of thallus crushed underfoot or torn away by wind may blow about or travel attached to some animal and later become deposited in a suitable habitat. Of the special asexual reproductive structures in lichens, small groups of algal cells enveloped in wefts of fungal hyphae (*soredia*; Fig. 1–1b) are the most commonly encountered. Soredia form either as a diffuse powdery layer covering the whole surface of the thallus or in special clearly delimited structures (*soralia*) which often have characteristic shapes and locations on the thallus (e.g. only on lobe-ends, only on the surface). *Isidia* (Fig. 1–1b) differ from soredia in that instead of being powder-like they are cylindrical, globose, coral-like or scale-like organs arising from the surface of the thallus as outgrowths from the cortex. Isidia consequently consist of a layer of fungal tissue enclosing many algal cells. Both soredia and isidia are dispersed by wind, rain-splash, rain-trickles, browsing invertebrates, birds and so on.

1.6 Growth

One of the best known features of lichens is their very slow growth rate and considerable longevity. Growth in crustose and foliose species is mainly marginal, so that rosette-like plants are commonly produced, while fruticose species mainly grow apically so that their height (erect species) or length (pendent species) increases with age. The annual radial increments of British crustose and foliose species are mainly in the order of 0.5–5.0 mm, while fruiticose species usually grow more quickly and some may achieve 1–2 cm increases in height or length each year. Plants several centuries old are commonly encountered in Britain and ages of up to 4500 years have been claimed for the crustose *Rhizocarpon geographicum* in Greenland. Growth rates vary even within the same species, however, depending on the environmental conditions.

With increasing age the central parts of some foliose species die and disintegrate while their margins are still actively spreading outwards so that rings or crescents are produced. The bare central areas frequently become colonized by new plants of the same species so that concentric thalli can be formed.

A slow growth rate does not mean that a species cannot provide an extensive cover on a substrate in a fairly short time. On concrete, for example, many separate individuals of the crustose *Lecanora dispersa* (Fig. A1–2a) can achieve an 85% cover in five years. A few species may have optimum sizes, after which death ensues, but in most cases the life of a lichen is determined by the longevity of its substrate (e.g. the tree on which it occurs) or competition from other plants (including other lichens) rather than any inherent ageing process in the thallus itself. Signs of lichen death are discussed in Section 2.3.

1.7 Physiology and synthesis

In recent years rapid progress has been made in the study of lichen physiology, both in terms of responses to environmental factors and chemical processes occuring within the thalli themselves. Many species occur on exposed surfaces where the supply of water and nutrients is sporadic, and consequently they may have to withstand periods of desiccation when no water is available either as rain or dew. When conditions are moist, respiration and photosynthesis occur rapidly but when the thalli dry out these and other processes come to a halt. As lichens lack a protective waterproof cuticle they are able to respond rapidly to changes in the moisture content of their environment and to take in water (and dissolved substances) over the whole thallus surface. Different species are adapted to specific environmental conditions with respect to their water relations. They not only withstand extremely high and low temperatures and drought conditions but actually need fluctuations in these factors to survive.

When the lichen is moist the cells of its algal partner carry out photosynthesis and produce excess carbohydrates which pass through their cell walls and into the fungal hyphae in the form of mobile sugar alcohols (termed polyols; e.g. ribitol) or glucose. The nature of the mobile carbohydrate depends on the genus of algae concerned. Inside the fungal hyphae other polyols (e.g. mannitol) are then produced and used to build up new fungal tissue as required.

Mineral nutrients are in most cases derived from ions dissolved in rain water, but in some instances there are indications that some substances may be taken up from the substrate. Many metals in particular can be accumulated to very high concentrations within the thalli (Section 3.4).

Where the lichen has a blue-green phycobiont, the blue-green algal cells are able to fix atmospheric nitrogen which then passes into the fungal hyphae. Species lacking blue-green algae presumably obtain the nitrogen they require from nitrates dissolved in rain water.

Many lichens produce one or more secondary metabolic products which are deposited as crystals on the outer surfaces of the fungal hyphae. The majority of these crystals are weak phenolic acids (e.g. depsides, depsidones) and are termed *lichen acids* because most of them are unknown in other groups of organisms. Many lichen acids are colourless and of uncertain function but a few more brightly coloured ones (e.g. usnic acid) are produced most abundantly in well-lit situations and may, when in the cortex, serve to protect the algae from excessive illumination: brown pigments in *Parmelia* and the orange anthraquinone parietin in *Caloplaca* and *Xanthoria* may have a similar rôle. The characteristic colours produced by some lichen acids with simple reagents are a valuable aid in identification (Section A1.4).

The physiological processes involved in the production of lichen thalli from separate germinating ascospores and algal cells in nature are unknown. In the laboratory, numerous attempts to synthesize lichens from their isolated components have been made, but only in one instance (first reported in 1877 and then repeated in 1967 and 1970 for *Endocarpon pusillum*), have the isolated elements produced a thallus resembling in all respects that encountered in nature. The key to successful synthesis appears to rest in the duplication of the field environmental conditions; something notoriously difficult to achieve in the laboratory. The components of a lichen can consequently perhaps be viewed as two organisms united in adversity.

1.8 Ecology and distribution

Lichens occupy a wide range of habitats from tidal zones on rocky shores to mountain summits, from hot deserts to the arctic and antarctic, and from tree leaves to tree trunks. A few species occur in a very wide range of habitats over several continents, but most tend to be restricted to particular ecological niches and to have characteristic distributions.

Both the chemical nature and the texture of the substrate and microclimatic variables are important limiting factors. On a single tree one can find assemblages of different lichen species on different parts, related to conditions prevailing in each particular microhabitat. The acidity of the bark of a tree appears to be of particular importance; birch and pine have more acid barks than oak and so tend to support communities adapted to more acidic conditions. If the acidity of the bark is changed by any factor the lichens present may also change (Sections 2.4; 3.6; 5.4). Most lichens occurring on trees (*corticolous* species) seem to be bark-type, rather than tree-species, characteristic. Lichen hyphae only penetrate the outer dead layers of bark and so do not damage their hosts. On barkless wood the lichens present (*lignicolous* species) tend to be ones normally favouring acid bark. Species on rock (*saxicolous* species) vary in a similar way to those on trees; particularly marked differences occur between species in communities on hard limestones and coarse-grained sandstones (something readily observed on tombstones of different rock types); some saxicolous species occur on man-made substrates (Section 4.4). Species on the ground (*terricolous* species) can form dense swards under suitable conditions; the tundra 'lichen-heaths' of the northern boreal and subarctic regions resemble to some extent those on some Scottish mountain tops. Peaty soils support quite different assemblages from calcareous grasslands and sand dunes.

Lichens are often stated to be primary colonizers of substrates (such as rock) but there is little evidence that they initiate a succession leading to vascular plants. In general they have poor competitive abilities compared with mosses and flowering plants and grow most luxuriantly in sites not favourable to higher plants. Within lichen communities succession and competition occur; faster growing fruticose species often tend to give way later to crustose communities on some types of rock.

In a particular climatic region each microhabitat on each substrate tends to acquire a particular assemblage of species if not affected by man. These communities are often so distinctive that they are given latinized names (e.g. *Lobarion*, *Xanthorion*) based on their characteristic species.

Although dual organisms, lichens nevertheless have distinctive distributional types comparable to those seen in flowering plants. Even within the British Isles many species show distinctive types of distribution (Section 4.7) related to a wide variety of factors (substrate limitations and climatic). On a world scale some lichens (e.g. *Parmelia sulcata*) have an extraordinarily wide distribution, but most tend to follow major vegetational zones and subzones. Genera, subgenera and sections often have marked centres of diversity, just as do the flowering plants. Relatively few species are restricted (*endemic*) to small geographical regions. Perhaps only two or three are really endemic to Britain.

2 Sulphur dioxide

2.1 Sulphur accumulation

Analyses of the sulphur content of specimens of a species in areas affected by sulphur dioxide pollution show that, as the most polluted area is approached, their sulphur content rises markedly (Table 1). Lichens accumulate sulphur from dilute solution just as they do metals (Section 3.4), but little is known of the rate at which sulphur is taken up in the field. Live thalli can accumulate much more sulphur than dead thalli (six times more in *Usnea filipendula*), indicating that a combination of active and passive processes is involved in uptake. Most uptake occurs under moist environmental conditions (i.e. when the thalli are most active physiologically).

Table 1 Relationship between total sulphur content (p.p.m. dry weight) and sulphur dioxide levels in the air in three corticolous species in West Central Scotland (based on data provided by Dr G. P. O'Hare)

Species	Mean winter sulphur dioxide levels (approx.) in $\mu g/m^3$				
	<30	35	40–50	55	60–70
Evernia prunastri	382	589	794	1129	—
Hypogymnia physodes	537	—	545	—	1509
Usnea subfloridana	254	676	1101	—	—

Sulphur in the air enters thalli both in solution (as sulphate, sulphite, and bisulphite ions or sulphurous acid; their frequencies of occurrence depending on the pH) and in gaseous form (sulphur dioxide and sulphur trioxide). Translocation within the thallus and sites of accumulation are obscure but some sulphur clearly enters the living protoplasts of the partners because basic metabolic processes are disrupted (Section 2.2). The behaviour of sulphur compounds in the thalli may vary in different species and be important with respect to their resistance to sulphur dioxide pollution (Section 2.5).

2.2 Effect on photosynthesis and respiration

The effect of sulphur dioxide on photosynthesis in the algal cells is probably the main factor responsible for the sensitivity of lichens to this pollutant. Photosynthetic processes are disrupted according to the concentration of sulphur dioxide with, progressively, a temporary reduction in rate and subsequent recovery, a permanent reduction in rate without chlorophyll breakdown, and a permanent reduction in rate related to chlorophyll breakdown. Damage is most severe under moist conditions at very low pH values; at pH 3.2–4.4 chlorophyll can be irreversibly oxidized (I) and at pH 2–3 is converted to phaeophytin (II) or broken down even further.

(I) chlorophyll a $\rightarrow$. chlorophyll a^+
(II) chlorophyll $a + 2H^+$ $\rightarrow$ phaeophytin $a + Mg^{++}$

At higher pH values chlorophyll is not degraded except at unusually high sulphur dioxide concentrations (when interference with electron transport chains is to be expected). In flowering plants also sulphur dioxide acts on photosynthesis by causing a disorientation in the chloroplast membranes, so reducing ATP production and inhibiting the enzyme carboxylase; the same may well occur in lichen algae. The greater susceptibility of lichens to this pollutant may be partly due to them having a much lower proportion of chlorophyll to total tissue. Experiments involving sulphur dioxide in solution show that different species are affected to different extents by the same pollutant levels; in general the order of sensitivity correlates with that expected from field observations (Fig. 2–1). Fumigation experiments tend to provide better correlations

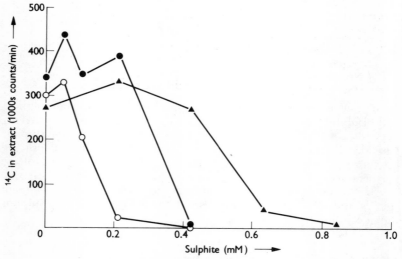

Fig. 2–1 Effect of sulphite on ^{14}C incorporation in *Usnea subfloridana* (o), *Hypogymnia physodes* (●) and *Lecanora conizaeoides* (▲) (from HILL, D. J., *New Phytol*, **70**, 831–836, 1971).

with field data, perhaps as artificial saturation overcomes some resistance mechanisms (Section 2.5). Dry thalli are much less sensitive than moist thalli and the rate at which the gas flows over thalli may also affect the amount of damage caused.

Unlike photosynthesis, respiration is a function of both the fungal and the algal partners. The separation of these components in experimental work is almost impossible, but from studies on non-lichenized fungi (Section 7.2) the fungal partner alone would be expected to show some adverse effects. Experiments involving sulphur dioxide in solution show that, sometimes after an initial increase in rate, all species examined have their respiration rates decreased at levels of the pollutant found in nature. In the case of both photosynthesis and respiration, sulphurous acid and bisulphite ions appear to be more toxic to lichens than sulphite, which in turn is more toxic than sulphate. The drastic effects of sulphur dioxide can also be studied in transplanted material; in one experiment specimens of *Ramalina farinacea* transplanted into Newcastle upon Tyne lost 84% of their chlorophyll and had their respiration rate reduced to 20% of normal in five weeks.

2.3 Effect on vitality

Some of the metabolic effects of sulphur dioxide lead to injury symptoms. When a lichen is killed (by pollution or natural factors) the symptoms include bleaching (chlorosis), a tendency to peel away from the substrate, and colour changes due to the breakdown of lichen acids (e.g. formation of reddish compounds in the medulla of *Parmelia sulcata*). In foliose species the centres often die first to leave arcs of marginal lobes which may continue to grow. Growth rates may also be reduced; in the West Riding *Parmelia saxatilis* in polluted sites has been recorded as showing a 32% reduction in marginal advance compared with controls over a three-year period.

A survey of the sizes of thalli of fruticose lichens on trees along a transect from an unaffected into a polluted area shows that their sizes decrease, normally conspicuous species being reduced to minute fragments which are easily overlooked (Fig. 6–4). Ascocarps in many of the larger lichens are normally found on mature (large) plants and so, not surprisingly, become very rare on them under pollution stress. In one Canadian study while 100% of specimens of *Physcia millegrana* were fertile in unaffected sites, only 3.3% had apothecia at its innermost stations in Montreal. Even when ascocarps are formed, the spores may show decreases in percentage germination. A few crustose species, in contrast, commonly have apothecia in urban areas (e.g. *Lecanora dispersa*). Many of the foliose and fruticose species used in pollution surveys reproduce by asexual methods, but even the production of these propagules can become reduced at their highest tolerated pollution levels; the viability of soredia is also depressed at levels of sulphur dioxide corresponding closely to the field sensitivities of the species from which they are derived.

In addition to changes in chlorophyll (Section 2.2) algal cells can exhibit a variety of effects, including increased plasmolysis (probably arising from damaged cell membranes), increased proportion of dead cells, and a reduction in the proportion of actively dividing cells. Nitrogen fixation by blue-green algae may also be markedly reduced; in one case of material transplanted into a town, by 80–90% in 3–4 weeks.

2.4 Effect on substrate preference

Rainfall in polluted areas becomes acidified (Table 6) and so tends to acidify substrates on which it falls; the extent to which substrates are affected depends on their buffer (neutralizing) capacity. Trees with acidic barks and low buffering capacities (e.g. birch) are the first to lose their lichens, while those with more alkaline barks and higher buffering capacities (e.g. elm) retain them longest. Under conditions of moderate pollution the bark of trees with intermediate bark characteristics (e.g. oak, sycamore) becomes acidified and supports distinctive assemblages of species (Section 6.3). Limestone tombstones and asbestos-cement also exert a neutralizing effect; consequently communities on these are less affected by sulphur dioxide than those on more acid rocks (e.g. sandstones). In heavily polluted areas some species normally found on acidic rocks start to grow on asbestos-cement when this has become sufficiently acidified (e.g. *Bacidia umbrina*, *Lecanora muralis*, Table 6). As variations in pH can occur even on different parts of trees and walls, some species become confined to nutrient rich wound-tracks (high pH) on trees and near basic run-off (e.g. from adjacent mortar) on sandstone walls at their nearest stations to pollution sources.

2.5 Mechanisms of resistance

Whether a species can persist in a particular site under sulphur dioxide stress depends on a multiplicity of factors varying from species to species and habitat to habitat. These can be divided into two main categories: 'avoidance' and 'tolerance' (Fig. 2–2). Avoidance is achieved by reducing the amount of pollutant the lichen takes in; wettability of the thallus is particularly important here. The very tolerant *Lecanora conizaeoides* produces large amounts of fumarprotocetraric acid, making it almost unwettable in the natural state. When artificially saturated it shows a respiratory sensitivity not unlike that of much less tolerant species. Surface area, physical barriers (e.g. cortex structure), and buffer capacities of the substrate and the internal tissues of the thallus may all contribute to avoidance mechanisms. Tolerance (protoplasmic resistance) is dependent on the state of maturity, vitality, and the pH,

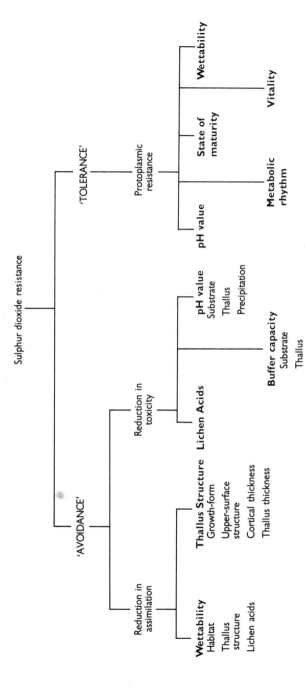

Fig. 2-2 Possible methods of resistance to sulphur dioxide pollution in lichens (adapted from TÜRK, R., WIRTH, V. and LANGE, O.L., *Oecologia* **15**, 33–64, 1974).

buffering capacity, wettability and permeability of the cells themselves.

Material of the same species collected from different regions and habitats varies with respect to some physiological parameters (e.g. optimum degree of saturation for maximum photosynthesis). Whether genetic adaptation is involved is unclear at the present time, but there are indications that more pollution-resistant races may arise in some species occasionally.

The complexity of resistance phenomena in lichens (Fig. 2–2) may explain why fumigation experiments and visual estimates of damage tend to correlate better with field observations than short-term measurements of single physiological responses in more artificial experimental situations.

3 Other Pollutants

3.1 Smoke

Smoke is the most familiar air pollutant because it is easily seen, but it is now clear that smoke itself (i.e. particulate carbon compounds) is much less toxic to lichens and vascular plants than invisible gaseous pollutants such as sulphur dioxide. High smoke and sulphur dioxide levels often occur together in urban areas and so are difficult to study in isolation in towns; investigations in rurally sited industrial areas with little or no smoke have been compared with lichen surveys in areas affected by both and show that sulphur dioxide rather than smoke limits the lichen vegetation. Indeed, there is no certain evidence that smoke affects lichens adversely. A comparison of the lichen vegetation with mean levels of both pollutants indicates that smoke is of little importance (Table 2).

Table 2 Comparison of mean winter 1967–70 smoke and sulphur dioxide levels and the lichen vegetation at selected sites

Station	Smoke ($\mu g/m^3$)	Sulphur dioxide ($\mu g/m^3$)	Lichen zone(s) (Table 5)
Leicester 14	89	175	0–1
Kew 1	38	150	2–3
Buxton 2	17	126	3
Sheffield 60	43	88	3
Dursley 5	22	87	3–4
Hayfield 2	102	84	3–4
Plymouth 13	97	82	3–4
Prestwood 1	32	60	6
Didcot 6	22	39	7
Torquay 3	33	32	8
Llanberis 1	29	27	9

Soot (i.e. accumulations of smoke particles) is rather alkaline and able to decrease the acidity of substrates on which it settles. Close to city centres the pH of bark often rises slightly as a result of this effect; the abundance of the bright emerald-green alga *Pleurococcus viridis* on slightly shaded trees in many British cities may be due largely to bark enrichment by soot. If soot deposition is extreme some lichens able to tolerate the co-existing sulphur dioxide levels accumulate soot particles amongst their hyphae and so develop blackened thalli; this is particularly marked in *Lecanora dispersa* which has a whitish thallus in soot-free areas but forms unsightly black stains on concrete in the centre of London and other cities.

3.2 Fluorides

Fluoride pollution arises mainly from aluminium smelters and some brickworks. In the case of brickworks, high sulphur dioxide levels frequently occur as well, but with aluminium smelters fluoride is often the only pollutant. Usually emitted either as hydrogen fluoride gas or particulate pollutants, fluorides react very easily with other substances and tend to be washed out of the air or otherwise deposited fairly close to their source. Fluorides disrupt many processes in plants, causing severe damage to herbaceous plants, mosses and trees; the latter may be killed and are thus absent in severely affected sites. Fluorosis (weakened and deformed bones) in cattle is also a problem in areas with very high fluoride levels. Transplants with lichens show that fluorides cause chlorosis, necrosis and thallus disintegration; the extent of the visible damage is related to the accumulation of fluorides in the tissues. Most lichens are eliminated by high levels of fluoride but, as is the case with sulphur dioxide, some species are more sensitive than others and so a zonation pattern develops around the source. The order of sensitivity may differ in some instances from that found in areas affected by sulphur dioxide pollution. Little is known of the precise levels of fluorides eliminating different lichen species, but in the laboratory even $4 \mu g/m^3$ of hydrogen fluoride for nine days can cause visible injury to some species.

Studies of both zonation patterns around fluoride sources and the fluoride content of lichen thalli reveal the pattern of contamination in affected areas. The severely affected region tends to be in close proximity to the source; in the case of the Fort William smelter in Scotland, the affected area of lichen vegetation is restricted to an ellipse about 4 by 1 km (the size and shape depend both on topography and the height of the chimneys). In the British Isles fluoride pollution is local, but particularly important in the south Bedfordshire, north Buckinghamshire and Peterborough brickfields, and on a far smaller scale in the vicinity of aluminium smelters such as those at Fort William, Holyhead, Invergordon and Kinlochleven.

3.3 Car fumes, hydrocarbons and dust

Cars emit carbon monoxide, oxides of nitrogen, hydrocarbons, lead-containing compounds, and other substances. Under suitable climatic conditions in urban areas with intensive car traffic, *photochemical smog* can develop, consisting of ozone, peroxyacetyl nitrate (PAN), nitrogen oxides and other compounds. This is a particularly severe problem in California, where many vascular plants are adversely affected, but these compounds do not seem to form in sufficient concentrations to cause comparable problems in the moister climate of the British Isles. It should be noted that the word *smog*, as formerly used in Britain, was applied to concomitant high smoke and sulphur dioxide levels and so is not to be confused with photochemical smog. The effects of the constituents of photochemical smog on lichens appear to be minimal.

The impact of car fumes in Britain can be assessed by studying the lichens present on trees by major roads in areas with little sulphur dioxide pollution. Such investigations indicate that traffic fumes have little effect on roadside lichens except where roadside trees are closed in by steep banks which restrict air flow and so tend to inhibit pollutant dispersal. Where roadside trees are almost completely devoid of lichens in sites little affected by sulphur dioxide, rock salt (used extensively to free roads from ice and snow) splashed up by the traffic may be the important inhibitory factor.

The lead content of lichens near roads has not been studied as intensively as has that of mosses and vascular plants. Investigations in the U.S.A. and Britain, however, show that the lead content of lichens increases following the construction of new roadways near them, and that the concentration in a particular species is greatest close to the road, and falls rapidly with distance away from it. There is no evidence that lead pollution leads to the loss of any lichens from roadside habitats; indeed, a few species preferring rocks rich in heavy metals (e.g. *Stereocaulon pileatum*) have become widespread on sandstone roadside walls over the last few decades.

Hydrocarbons, to judge from detailed investigations around the Fawley Oil Refinery in Hampshire, appear to have little or no effect on lichens, and this is probably true for hydrocarbons originating from car exhausts.

Dust enriched with horse manure was commonly splashed or blown on to roadside trees until about 60 years ago leading to a reduction in bark acidity which encouraged lichen assemblages characteristic of basic barked trees (i.e. *Xanthorion*). The increase in the metalling (surfacing) of roads and the gradual disappearance of horse-drawn traffic have contributed to the decline of such communities in Britain. *Xanthorion*, lacking its more demanding species, is, nevertheless, commonly encountered on roadside trees today where sulphur dioxide levels permit its development. Dust from limestone quarries and cement and lime works is alkaline and has a neutralizing effect, tending to reduce bark acidity and promote the development of *Xanthorion*: some species normally confined to limestone rock may also appear on trees (e.g. *Physcia caesia*). If there is so much dust that a white film covers all the surrounding vegetation, most corticolous lichens are lost; around major limestone quarries lichen-depleted, *Xanthorion*, and unaffected zones can thus sometimes be distinguished.

3.4 Metals

Some lichens growing on rocks rich in heavy metals (e.g. iron, lead, zinc) accumulate these to high concentrations in their thalli; some species (e.g. *Lecidea macrocarpa*) assume rusty colours when growing on rocks rich in iron. To what extent individual species may have requirements for particular metals is unknown; one species, confined to vineyard poles subject to copper sprays, can contain 5000 p.p.m. of copper (*Lecanora vinetorum*). Under conditions of metal pollution, lichens also take up metals. As one proceeds away from a metal smelter the metal contents of lichens fall (Table 3); such surveys provide an indication of both the

Table 3 Metal content (p.p.m. dry weight) of *Peltigera rufescens* along a 1 km transect at Risby Warren (Lincolnshire) in an area subjected to metal contamination from steel works (data from SEWARD, M. R. D., *Lichenologist* 5, 423–433, 1973).

Metal	Site						'Control'
	(a)	(b)	(c)	(d)	(e)	(f)	
Chromium	127	64	61	33	42	25	26
Copper	91	54	34	20	27	20	16
Iron	90 380	77 110	32 020	13 760	26 820	15 170	14 150
Lead	454	139	125	46	120	59	79
Manganese	5 000	3 293	747	371	838	386	372
Nickel	38	24	33	11	52	26	10

Note: Site (a) is nearest to the steel works and (f) furthest from it; the 'control' is material collected in the area in 1907.

pattern and the extent of heavy metal contamination in an area. As emissions from smelters include high levels of sulphur dioxide in addition to particulate metal-containing pollutants the effects of metals in isolation are difficult to assess in such areas. Water run-off on roofs with lead, zinc or copper flashings, areas beneath copper-containing power lines and below window frames and barbed wire, provide extreme situations and lichens may be almost entirely eliminated. A few common species are, nevertheless, able to grow directly on iron (e.g. *Lecanora polytropa*).

Laboratory experiments show that different species show selective uptake, absorbing some metal ions in preference to others. Once inside the thallus they are deposited either on the outer surfaces of the walls of fungal hyphae or within the walls themselves; positions where they may have relatively little effect on the cell protoplasts as the concentration inside the cells will be much less than that in or on their walls. A few metal ions (e.g. potassium) may be accumulated within the cells themselves; possibly bound to proteins. Uptake itself appears to be mainly passive (by diffusion), but in *Cladonia* the tendency for metals to be concentrated in the uppermost living portions of the thallus indicates active processes may also play some part.

Lichens accumulate heavy metals to exceedingly high concentrations around smelters but nevertheless survive. In sites subject to metal-rich run-off (see above) where lichens are lost, some metals, notably copper and zinc, appear more toxic than others, such as iron. Copper-containing sprays have been recommended for the removal of lichens from trees. The mechanisms of metal toxicity in lichens are obscure, but in one experiment copper caused a marked reduction in respiration rate, while other metals tested (including zinc) had either less severe or no effects.

3.5 Radionuclides and radiation

Radionuclides (radioactive isotopes of metals) are accumulated in lichens in a similar way to other metals (Section 3.4), but are a cause of some concern because of the ionizing radiation they emit. Radionuclides arising from nuclear detonations are dispersed in the upper atmosphere and finally deposited ('fallout') gradually over very wide areas. In the subarctic tundra, lichens are eaten by caribou and reindeer, which in turn are consumed by man; radionuclides taken in by lichens are concentrated at each step in the lichen-reindeer-man food-chain and in the late 1950s and early 1960s compounds such as caesium-137 and strontium-90 began to occur to disturbingly high levels in man. Since the general cessation of nuclear testing the levels in lichens, and so man, have tended to fall. The value of lichens as monitors of radionuclide fallout is therefore clear, and regular surveys continue to be carried out.

In contrast to man and most other organisms, and also in contrast to their behaviour with other pollutants, lichens are remarkably resistant to ionizing radiation; this may be because radiation acts on genetic mechanisms, processes presumably occurring at an extremely slow rate in lichens. Some species can survive within 6–8 m of a source of 1000 rad/day and it is estimated that at least some could tolerate 10 000–12 000 rad/day for three years; in one case, however, an exposure to 2700 rad/day for a year led to some reduction in the diversification of communities, but many species were still present.

3.6 Agricultural chemicals

Many chemicals are used in modern agriculture which have different effects on lichens. Copper-containing sprays (e.g. Bordeaux Mixture) used to control fungal diseases in orchards also result in the loss of most lichens (Section 3.4); increasing orchard hygiene has certainly contributed to the dramatic decline of a number of orchard-loving species (e.g. *Teloschistes flavicans*) over the last century. Precise data on the effects of most herbicidal and pesticidal sprays on lichens are not available, but field observations indicate that they can be harmful—in some instances eliminating all lichens. If applied from the air or blown by wind, lichens in adjoining areas may also be affected; this phenomenon may contribute to the paucity of species in intensively farmed flat areas of East Anglia.

Moderate doses of fertilizers such as superphosphate enrich the base status of bark and so promote species of the *Xanthorion* characteristic of normally nutrient-rich bark; trees in old grazed parkland are now one of the last strongholds of this community in Britain (Section 3.3). Excessive applications, in contrast, lead to hypertrophication which encourages the growth of algae such as *Pleurococcus* to such an extent that any lichens present may be smothered and eliminated.

3.7 Freshwater and maritime pollution

Compared with the information available on the effects of freshwater pollution on algae, fish and invertebrates, the effects of water pollution on lichens remain largely unknown. Those lichens characteristic of submerged and inundated rocks in fast-flowing rivers and lake margins are now rarely encountered where streams flow through intensively farmed land or urban areas. While this suggests effluents may be exerting some deleterious rôle, the situation appears to be complex and is compounded by disturbance arising from 'cleaning-up' operations which lead to deleterious increased turbidity (muddiness).

In the case of oil spills, oil itself exerts a smothering effect and many species undergo colour changes comparable to those seen when death occurs from other causes (Section 2.3); even crustose species tend to curl away from the substrate and are lost. Emulsifiers used to disperse oil are more harmful to lichens than oil alone, but their effects are decreased when mixed with seawater or oil and as the mixtures age. One component of the emulsifier BP 1002 appears to act by affecting the permeability of algal cell membranes within the lichen. Following severe oil spills and subsequent attempts to remove the oil, rocks can be left almost entirely devoid of live thalli. As lichens grow slowly (Section 1.6) recovery is also very slow and colonization may not be apparent for several years. Furthermore, investigations following the *Torrey Canyon* disaster of 1967 in the English Channel indicated that emulsifiers kill limpets and other herbivores which normally devour various algae; because of this, luxuriant growths of *Ulva* and other free-living algae can become firmly established on the rocks and prevent the regrowth of damaged lichens.

4 Other Factors affecting Lichen Distribution

4.1 Introduction

While various pollutants (Chapters 2, 3) appear responsible for most of the more spectacular changes in lichen distribution within the British Isles (Chapter 6), many other factors also lead to changes. If a particular habitat has been destroyed or is naturally absent in an area, corresponding lacunae will appear on distribution maps; the scarcity of trees in the Fenland basin (East Anglia), for example, is clearly shown on maps of many corticolous species. This, and less obvious factors discussed below, must be borne in mind in surveys aimed at ascertaining patterns of a particular pollutant's effects at both the national and more local levels; if they are not, misleading conclusions could be drawn.

4.2 Woodland management

Woodland management exerts profound effects on species characteristic of ancient woodlands. Studies of extant relics of ancient forests, such as the New Forest, in Hampshire, indicate that undisturbed primary hardwood forests in Britain probably had about 120–150 corticolous species per square km. In most woodlands in lowland Britain today, even in relatively unpolluted areas, this figure is often less than 50. The phenomenon correlates with the past management of the woodlands; the mature oak forest of Dalegarth Wood, in Eskdale, for example, with only about 68 species per square km, was clear-felled in the late eighteenth century, and its flora contrasts with other woodlands in Lakeland which have never been clear-felled and support over 100 lichen species per square km. Coppicing leads to a rapid drying out of the habitat, making it unfavourable to species requiring a humid environment; at Bradfield Woods in Suffolk where coppicing has been carried out since the thirteenth century or earlier, only 14 species per square km occur. Increased drainage may accentuate these effects. In the replacement of deciduous woodlands by coniferous plantations the characteristic lichen flora of deciduous trees is lost and the introduced conifers support very few species; the rich lichen flora of ancient pine woods in Scotland seems unable to migrate further south partly due to their remoteness and partly due to climatic factors (Section 4.7).

Some lichens, most of which belong to a community termed the *Lobarion*, are particularly sensitive to the disturbance of the microclimate in old forest areas. These 'old forest indicator' species are healthy and able to replace themselves in extant remnants of ancient forests but are relics in being unable to colonize more recently established woods. It is conceivable that the slight levels of sulphur dioxide now present in even the most rural parts of north-west Europe may interact with disturbance factors to reduce dispersal and establishment. Examples of 'old forest indicator' species in lowland Britain are *Lobaria* species (Fig. 6–1), *Nephroma laevigatum*, *Peltigera horizontalis*, *Sticta* species and *Thelotrema lepadinum*.

4.3 Heathland management

In lowland Britain heathlands dominated by ericaceous shrubs originate from forest clearance started in the Bronze Age. These are now an important habitat for terricolous lichens but if left undisturbed such heathlands tend to revert to forest and their persistence has depended on human management by grazing, cutting or burning (accidental or not). Burning of the northern England grouse moors at 10–15 year intervals is too frequent to permit the development of larger fruticose species (e.g. *Cladonia impexa*). A few heathland lichens (e.g. *Cetraria islandica*) are now relics in lowland Britain as a result of such effects, and also of increased drainage. Reduction in rabbit populations as a result of myxomatosis has enabled taller grasses to enter areas in which they could not previously survive, and threaten the persistence of some lichen species confined in Britain to the calcareous heaths in Breckland (East Anglia). Decreases in grazing pressure also affect communities of open chalk and limestone grasslands.

4.4 Man-made substrates

Many species originally restricted to natural rock outcrops in upland areas of Britain are now common throughout lowland Britain on man-made substrates such as asbestos-cement, bricks, concrete, tombstones and walls. Species of both calcareous and siliceous rocks have been able to take advantage of these new substrates (Table 4). Some species are almost unknown in Britain except on man-made calcareous substrates (e.g. *Caloplaca teicholyta*, *Candelariella medians*) where they are common (Fig. 6–2a), and active expansion in ranges of some species is occurring at the present time (e.g. *Xanthoria elegans* on concrete and tiles). That most man-made substrates tend to be rather alkaline is important as this exerts a neutralizing effect on sulphurous pollutants (Section 2.5) and enables lichens to penetrate further into towns on asbestos-cement and concrete in particular (e.g. *Lecanora dispersa*, *L. muralis*; Figs 5–2, 6–2c; see also Section 6.3).

The decreasing use of untreated timber may have led to the decline of a few lichens and the extinction of one species (*L. farinaria*) in lowland Britain; in contrast, fenceposts permit the extension of corticolous species into generally treeless areas (e.g. Shetland).

Table 4 Some species occurring on man-made saxicolous substrates in areas of lowland Britain where their natural substrates are absent

Acarospora fuscata s	*L. campestris* c	*P. nigricans* c
Bacidia umbrina s	*L. muralis* c, s	*Placynthium nigrum* c
Buellia aethalea s	*Lecidea lucida* s	*Protoblastenia rupestris* c
Caloplaca aurantia c	*L. sulphurea* s	*Rhizocarpon geographicum* c
C. heppiana c	*L. tumida* s	*R. obscuratum* s
C. saxicola c	*Lecidella stigmatea* c	*Rinodina subexigua* c
C. teicholyta c	*Ochrolechia parella* s	*Sarcogyne simplex* s
Candelariella aurella c	*Opegrapha chevallieri* c	*Solenopsora candicans* c
C. medians c	*Parmelia mougeotii* s	*Toninia aromatica* c
Lecanora atra s	*P. verruculifera* s	*Verrucaria sphinctrina* c
L. calcarea c	*Physcia caesia* c	*Xanthoria aureola*

c = mainly calcareous substrates; s = mainly siliceous substrates.

Man also causes changes in communities developed on a particular substrate. In addition to the effects of agricultural chemicals (Section 3.6) on trees in farmland, animal dung, urine and dust contribute to the development of *Xanthorion* communities characteristic of nutrient-rich (hypertrophiated) barks; such communities are dominated by species of *Physcia*, *Physconia*, *Ramalina* and *Xanthoria*. Changes in agriculture and roads (Section 3.3) have meant that such communities are often confined to rather isolated groups of trees; *Xanthorion*, like other communities formerly widespread on thatch, mud-capped walls and heathlands, is consequently tending to become relict (Section 6.5).

4.5 Drought

Drought in urban and some intensively drained agricultural areas undoubtedly contributes to changes in the distribution of some lichens. The concept that drought rather than toxic gaseous pollutants was entirely responsible for the absence of lichens in urban areas has now been discredited; the many arguments against this thesis include a few phenomena easily seen in the field: the absence of species from large rural areas around towns (including humid woodlands), the paucity of lichens in humid situations within them (e.g. islands in lakes), the presence of relatively rich lichen floras in towns with little sulphur dioxide pollution (e.g. Plymouth and Torquay), and the reduced lichen floras associated with isolated sulphur dioxide producing industrial plants in rural areas.

It is not easy to study the variables within the environment of a large urban area in isolation, and our knowledge of the effects of drought is still very inadequate. Increased aridity may favour the persistence of some species (Section 5.7), but others may be adversely affected.

4.6 Topography

Topography interacts in a striking way with air pollutants to control lichen distributions and so must always be taken into account in the interpretation of results obtained from lichen mapping studies. If the source is in a valley, temperature inversions tend to ensure that the valley is more polluted than the surrounding high ground; if, in contrast, the source is in an open situation, lichens on gently rising ground down-wind tend to be more affected than those close by in deep ravines. In such sheltered ravines lichen vegetation referable to higher zones (Table 5) may persist in marked contrast to the zones on the surrounding plateaux; thus 57 lichens persist on trees at Lullingstone Park (Kent), situated in a deep valley only 13 km south of the industrialized Thames Valley, while only five occur on trees in woodlands on the plateau 0.5 km south of this Park.

4.7 Climate

Climatic factors have profound effects on lichen distributions both on the national and the more local scales. Many species have distributions comparable to those seen in some of our geographically restricted flowering plants and, as in these, climatic factors are often the limiting parameters. Thus, for example, there are arctic-alpine lichens largely confined to high mountains in Scotland (e.g. *Cetraria nivalis*), ones almost confined to the extreme south coast where annual sunshine exceeds 1500 h (e.g. *Parmelia carporrhizans*), ones restricted to exceptionally high rainfall areas of the west (e.g. *P. laevigata*, *Pannaria rubiginosa*), and ones predominating in low rainfall areas of the east and south-east (e.g. *Anaptychia ciliaris*, *Parmelia acetabulum*). Apart from such rather extreme types, shown in Fig. 4–1, others become much rarer northwards (e.g. *Buellia canescens*, *Parmelia caperata*, *Rinodina roboris*) or southwards (e.g. *Alectoria fuscescens*, *Pseudevernia furfuracea*). The nature of the climatic factors involved varies from species to species and more than one may often be involved. Patterns due to air pollutants are superimposed on those determined by climate and for this reason some species are useful as pollution indicators in some, but not all, regions of the British Isles. *Parmelia caperata* (Fig. 6–1c), for example, is a useful indicator over most of lowland Britain, but its absence from large areas of north-east England, Scotland, and parts of central Wales, arises mainly from climatic factors.

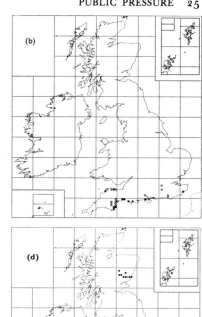

Fig. 4–1 Distribution types. (a) *Alectoria nigricans* (northern); (b) *Parmelia carporrhizans* (extreme southern); (c) *Pannaria rubiginosa* (western); (d) *Parmelia acetabulum* (eastern). o=pre-1960; ●=post-1960. Based on data from the British Lichen Society's Distribution Maps Scheme.

4.8 Public pressure

Recreational pressure is now severe in many habitats important for lichens. In addition to increased fire risks, damaging to both woodland and heathland communities, trampling is detrimental to terricolous species and climbing leads to saxicolous species being scoured off the rocks; indeed popular climbers' pitches are often easily identifiable by the absence of foliose lichens.

5 Mapping Air Pollution Patterns

5.1 Introduction

In 1859 L. H. Grindon attributed the declining lichen flora of South Lancashire to increasing air pollution and, in 1866, W. Nylander concluded from studies in Paris that lichens might serve as practical indicators of air quality. Subsequent studies scattered through over 400 scientific publications and dealing with areas in Europe, Japan, New Zealand, North and South America and the U.S.S.R. have vindicated Nylander's views. The more widely used of the various procedures employed in these investigations are summarized here.

5.2 Species mapping

Only through plotting the distribution of individual species on particular substrates can the relative sensitivities to pollutants of species in an area be ascertained; this is essential groundwork for the techniques discussed below. The closest sites of a species to the pollution source are of greatest value and results from such surveys are most clearly displayed as lines indicating the inner limits of species with different sensitivities (Fig. 5–1). As this method is dependent on the discovery and identification of often abnormal and minute fragments of specimens it is very time-consuming. Luxuriance and percentage cover are not taken into account easily in this approach, except by producing complex maps which are not suited to rapid interpretation. Particular attention has to be paid to the standardization of substrates and habitats, and for these and the above reasons this method is difficult for the non-specialist to employ.

Studies based on the number of species on particular substrates and the percentage cover of particular species or life-forms are also instructive. Such data can be plotted on maps or shown as graphs based on information from transects (narrow belts) out from the pollution source (Fig. 5–2).

5.3 Zone mapping

In 1912 it was appreciated that the lichen vegetation on trees in urban areas could be divided into zones easily recognizable in the field. Three or four zones were most commonly distinguished at first: (1) an inner 'lichen desert' with no lichens, or at least no foliose and fruticose species, (2) an intermediate 'struggle' or 'transition' zone where foliose and fruticose

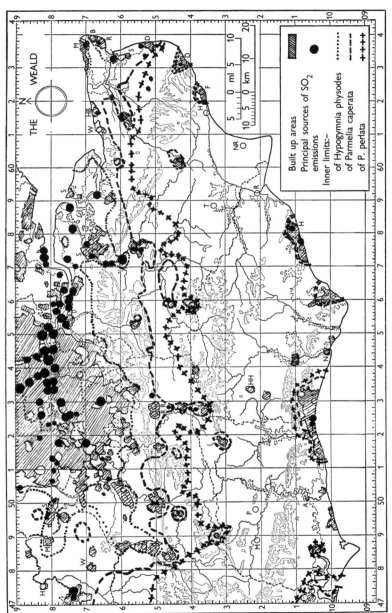

Fig. 5–1 Inner limits of three lichen species in south-east England; the sizes of dots indicating sulphur dioxide emission sources are proportional to the size of the source (from Rose, F., *Your Environment* **1**, 185–189, 1970).

species begin to appear but are poorly developed (often divided into an 'inner' and an 'outer' zone where foliose and fruticose species, respectively, first appear), and (3) an outer 'normal' zone with lichen

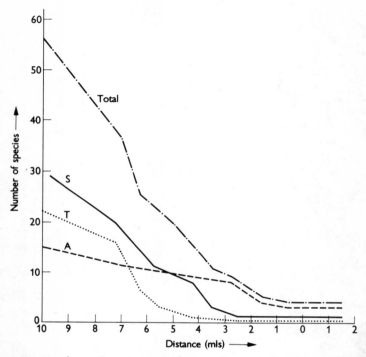

Fig. 5-2 Numbers of species occurring on asbestos-cement (A), sandstone (S) and trees (T) along a transect through the centre of Newcastle upon Tyne (from GILBERT, O. L., *Symp. Br. ecol. Soc.*, **5**, 35–47, 1965).

vegetation unaffected by pollution. Maps showing zones defined in these general terms are easily constructed, but their information content is low in relation to the complexities of the present situation in many regions. More informative maps are now produced by making use of the luxuriance and sensitivities of a wide range of species.

On transects from heavily polluted to relatively unpolluted regions the abundance of species varies from point to point with different species predominating at different distances along them; the lichen vegetation thus tends to fall into a series of noda, an ecologically important type of variation masked in maps based on the occurrence of individual species. Zones based on these noda provide the basis of the most useful type of

§ 5.3 ZONE MAPPING

zone scales when species are carefully selected and habitats standardized (Section 5.5). Such zones also tend to provide good correlations with measured pollutant levels (Section 5.6). The first scale for the estimation of sulphur dioxide in the British Isles using lichens was published by Dr O. L. Gilbert in 1968. Subsequent studies have enabled Gilbert's scale to be refined and extended in the case of corticolous lichens (Table 5). In this

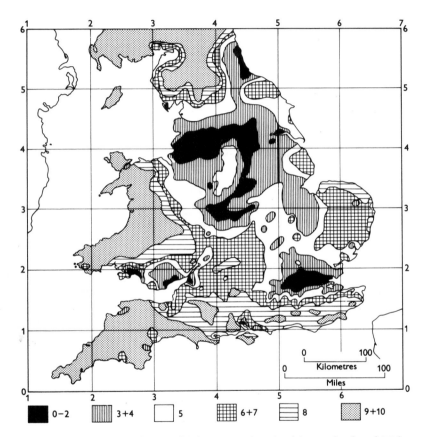

Fig. 5–3 Approximate limits of lichen zones (Table 5) in England and Wales. *Note.* (1) Local variations cannot be shown adequately at this scale and in the construction of boundaries emphasis has been placed on optimal sites, parts of most urban areas have zones lower than those indicated in regions mainly of zone 4 and above; (2) where boundaries are close intermediate zones have been omitted; and (3) differences between this map and the preliminary map in *Nature, Lond.*, **227**, 145–8, 1970, reflect more accurate information rather than any changes in zone boundaries.

Table 5 Zone scale for the estimation of mean winter sulphur dioxide levels in England and Wales using corticolous lichens (adapted from HAWKSWORTH, D. L. and ROSE, F., *Nature, Lond.*, 227, 145–148, 1970)

Zone	Moderately acid bark
0	Epiphytes absent
1	*Pleurococcus viridis* s.l. present but confined to the base
2	*Pleurococcus viridis* s.l. extends up the trunk; *Lecanora conizaeoides* present but confined to the bases
3	*Lecanora conizaeoides* extends up the trunk; *Lepraria incana* becomes frequent on the bases
4	*Hypogymnia physodes* and/or *Parmelia saxatilis*, or *P. sulcata* appear on the bases but do not extend up the trunks. *Lecidea scalaris*, *Lecanora expallens* and *Chaenotheca ferruginea* often present
5	*Hypogymnia physodes* or *P. saxatalis* extends up the trunk to 2.5 m or more; *P. glabratula*, *P. subrudecta*, *Parmeliopsis ambigua* and *Lecanora chlarotera* appear; *Calicium viride*, *Lepraria candelaris* and *Pertusaria amara* may occur; *Ramalina farinacea* and *Evernia prunastri* if present largely confined to the bases; *Platismatia glauca* may be present on horizontal branches
6	*P. caperata* present at least on the base; rich in species of *Pertusaria* (e.g., *P. albescens*, *P. hymenea*) and *Parmelia* (e.g., *P. revoluta* (except in NE), *P. tiliacea*, *P. exasperatula* (in N)); *Graphis elegans* appearing; *Pseudevernia furfuracea* and *Alectoria fuscescens* present in upland areas
7	*Parmelia caperata*, *P. revoluta* (except in NE), *P. tiliacea*, *P. exasperatula* (in N) extend up the trunk; *Usnea subfloridana*, *Pertusaria hemisphaerica*, *Rinodina roboris* (in S) and *Arthonia impolita* (in E) appear
8	*Usnea ceratina*, *Parmelia perlata* or *P. reticulata* (S and W) appear; *Rinodina roboris* extends up the trunk (in S); *Normandina pulchella* and *U. rubiginea* (in S) usually present
9	*Lobaria pulmonaria*, *L. amplissima*, *Pachyphiale cornea*, *Dimerella lutea*, or *Usnea florida* present; if these absent crustose flora well developed with often more than 25 species on larger well lit trees
10	*L. amplissima*, *L. scrobiculata*, *Sticta limbata*, *Pannaria* spp., *Usnea articulata*, *U. filipendulla* or *Teloschistes flavicans* present to locally abundant

Basic or nutrient-enriched bark	Mean winter SO_2 ($\mu g/m^3$)
Epiphytes absent	?
Pleurococcus viridis s.l. extends up the trunk	>170
Lecanora conizaeoides abundant; *L. expallens* occurs occasionally on the bases	about 150
Lecanora expallens and *Buellia punctata* abundant; *B. canescens* appears	About 125
Buellia canescens common; *Physcia adscendens* and *Xanthoria parietina* appear on the bases; *Physicia tribacia* appear in S	About 70
Physconia grisea, P. farrea, Buellia alboatra, Physcia orbicularis, P. tenella, Ramalina farinacea, Haematomma ochroleucum var. *porphyrium, Schismatomma decolorans, Xanthoria candelaria, Opegrapha varia* and *O. vulgata* appear; *Buellia canescens* and *X. parietina* common; *Parmelia acetabulum* appear in E	About 60
Pertusaria albescens, Physconia pulverulenta, Physciopsis adglutinata, Arthopyrenia gemmata, Caloplaca luteoalba, Xanthoria polycarpa and *Lecania cyrtella* appear; *Physconia grisea, Physcia orbicularis, Opegrapha varia* and *O. vulgata* became abundant	About 50
Physcia aipolia, Anaptychia ciliaris, Bacidia rubella, Ramalina fastigiata, Candelaria concolor and *Arthopyrenia biformis* appear	About 40
Physcia aipolia abundant; *Anaptychia ciliaris* occurs in fruit; *Parmelia perlata, P. reticulata* (in S and W), *Gyalecta flotowii, Ramalina obtusata, R. pollinaria* and *Desmazieria evernioides* appear	About 35
Ramalina calicaris, R. fraxinea, R. subfarinacea, Physcia leptalea, Caloplaca aurantiaca and *C. cerina* appear	Under 30
As 9	'Pure'

scale communities on moderately acid barks (e.g. oak) are separated from those on nutrient-rich or hypertrophiated (eutrophiated) barks (e.g. elm) enabling these differences also to be taken into account. This approach enables large areas to be surveyed rapidly by those familiar with the species involved (Fig. 5–3). Scales employing saxicolous habitats have also been constructed for higher pollution levels and these prove of particular value in urban areas where sulphur dioxide levels are high and trees scarce: *Lecanora muralis* is particularly valuable in this respect (Table 6).

Table 6 Biological scale for the status of *Lecanora muralis* compared with distance from the centre of Leeds, pH of rainwater and sulphur dioxide levels (based on data of Dr M. R. D. Seaward)

Distance from Leeds city centre in 1970 (miles)	Status of Lecanora muralis	Mean annual SO_2 ($\mu g/m^3$)	Rainwater pH
0–1.5	Absent	>240	4.4–4.7
1.5–2.5	On asbestos-cement tile roofs	200–240	4.7–4.9
2.5–3.5	On asbestos-cement tile and sheet roofs	170–200	4.9–5.1
3.5–5.5	On asbestos-cement roofs, cement, concrete and mortar	125–170	5.1–5.5
Over 5.5	On asbestos-cement roofs, cement, concrete and siliceous wall capstones	>125	Over 5.5

Zone scales have been criticized as being difficult to use because of the numbers of species which ideally should be employed. However, useful data can be obtained by the non-specialist by using somewhat simplified scales, as proved in a survey by school children, discussed in Appendix A2.4.

5.4 Other methods

Indices based on the species present can be calculated and plotted on maps. The most widely used of these is the Index of Atmospheric Purity (IAP) defined by the formula:

$$IAP = \sum_{n}^{1} \frac{(Q \times f)}{10}$$

[n = number of species at the site; f = frequency (cover) of the species; and Q = mean number of other species growing with the species in the area.]

Numerical methods are very time-consuming and do not appear to produce more accurate maps than zone scales. Studies of the contents of sulphur (Section 2.1) and metals (Section 3.4) of thalli also provide indications of pollutant patterns; where several pollutants are involved this method may enable their different patterns to be distinguished. Monitoring of changes in photosynthetic and respiratory rates (Section 2.2) may show when pollutants are starting to have adverse effects long before thalli die, and a recent study in Germany based on the activity of the enzyme phosphatase in one species enabled maps from this to be produced. Transplanted material also provides methods of determining pollutant patterns by measuring the accumulation of compounds in them, measuring changes in physiological parameters, and noting the time taken for thalli to die; material can be displayed on boards in treeless areas (or where all lichens have been lost) and maps based on rates of death, etc., constructed.

5.5 Indicator species and standardization

The most reliable and useful indicators are species which (1) would be expected to occur in the area were it not for pollution (i.e. where suitable substrates persist and the species used to occur), (2) still occur in adjacent areas (so that colonization would be feasible if pollutant levels were lower), (3) are known to be affected by the pollutants concerned, (4) are characteristic of open habitats, (5) show a range of sensitivities, and (6) are easily recognized in the field. Such criteria can be applied in regions where the lichen flora is well known, but for relatively little known countries considerable preliminary work is required. As the value of species may vary in different regions (Section 4.7), regional variations must be taken into account in scales designed for large geographical areas (Table 5).

Standardization procedures aim to eliminate factors other than air pollution from consideration. Trees of the same species or at least similar bark characteristics (supporting similar communities in unaffected areas) which are also vertical, in comparable situations (exposed, free-standing, not in ravines or dense woods) and of similar ages (e.g. 0.5–1 m diameter at breast height) are most useful. The separation of trees with moderately acid barks (e.g. oak) from naturally basic (e.g. elm) or nutrient-enriched (hypertrophiated) barks is also valuable (Table 5; horse-chestnut, sweet chestnut, plane, beech and conifers are not considered in this Table). Where possible several trees at a site should be examined and the mean point on the scale used. Where a survey aims at determining patterns of a particular pollutant, allowances for local sources of other pollutants must also be made (e.g. avoiding main roads and trees in pastures subject to spraying). The reasons for these various precautions will be seen from Chapters 3–4.

In the case of saxicolous substrates, man-made materials such as asbestos-cement prove particularly valuable (Table 6). Comparison of natural rocks present various problems due to variations in chemical composition and texture. When dealing with calcareous tombstones relict populations must also be considered (Section 6.5).

5.6 Correlation with pollutant levels

The comparison of the lichen vegetation adjacent to established air pollution recording gauges (Table 2) has enabled lichen zones to be calibrated (Tables 5 and 6). Correlations with mean winter values (i.e. in the periods when lichens are most active physiologically) agree to a remarkable extent and predictions of these levels based on the lichen vegetation agree with that derived from subsequently erected gauges; nevertheless, caution is required (Section 5.7). As levels of pollutants fluctuate from year to year, values averaged over the last 3–4 years provide the best correlations. In areas where recording gauges are absent, the acidity of rainwater, sulphate content of tree leaves and soil, and so on, have all been used as a guide to pollutant levels.

5.7 Interpretation and value

In most methods described above it is the lichens themselves which are being mapped: only in regions where correlations with pollutant levels have been established can lichen patterns be taken as estimates of levels of particular pollutants. By extrapolation from European studies, lichen zonation patterns developed in polluted areas may be assumed to be related to some pollutant(s), but they may not reflect the most harmful ones under some climatic conditions; in arid areas of the U.S.A. components of photochemical smog (Section 3.3) may reach levels harmful to crop plants in dry periods when lichens are not physiologically active (in such a case the lichens may show the mean sulphur dioxide pattern, but *not* that of the smog). Even within Europe the sensitivity of a particular species may show some variation according to its geographical position. In the oceanic climate of western Europe there is a tendency for species to be eliminated by lower sulphur dioxide levels than in some eastern European countries, something possibly related to the lengths of time for which the lichens are physiologically active in the year. A scale for the estimation of mean sulphur dioxide levels in one country cannot, therefore, be reliably employed in another without careful preliminary investigations.

Two other factors require more intensive investigation. First, while correlations with mean sulphur dioxide levels have been demonstrated, it is conceivable that short exposures to exceptionally high levels under humid conditions might eliminate some species (transplant work indicates this is not so for all). Secondly, while zonation patterns develop around new pollution sources within a few years, there are few data as to

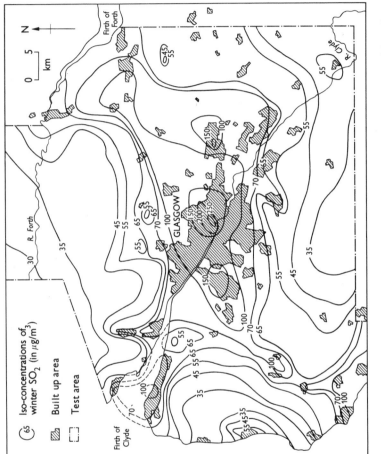

Fig. 5-4 Pattern of mean winter sulphur dioxide levels in west central Scotland based on data derived from both recording gauges and the lichen vegetation (from O'HARE, G. P., *Area*, **5**, 223–9, 1973).

how quickly recolonization may occur. Some recovery may start almost immediately, but such reductions in sulphur dioxide as have been achieved in urban areas of Britain (Section 7.1) have not been sufficient to expect marked improvements in our corticolous lichen flora, although increased frequencies of a few species has been recorded in some (e.g. *Lecanora conizaeoides*).

The main value of zone scales is that, where correlations with pollutant levels have been established, large areas can be surveyed quickly and maps produced which give a valuable indication of pollution patterns. A comparable amount of information could alternatively only be derived by having very many recording gauges operating for several years. Scales using corticolous lichens are of little value at very high mean sulphur dioxide levels (over 170 μg/m^3) and so are most suited to surveys in suburban and rural areas where in any case recording gauges are sparse and there are more trees. Data derived from the lichens and gauges can be satisfactorily combined to produce most useful maps (Fig. 5–4). Also, further studies of saxicolous species will certainly lead to scales covering higher levels in the future. As coniferous trees appear to be affected within the most sensitive part of the corticolous scales (Section 7.2), lichen scales can provide indications of areas suited to economic forestry; deteriorations of the lichen flora are also important in broader aspects of conservation (Section 7.5).

Even in areas where cause-and-effect are not established, the initiation of lichen zonation patterns should be a warning signal as air pollution is most likely to be the cause. It has also been suggested that biological scales are of particular value where many pollutants are involved as organisms will show their pattern, and possibly indicate adverse synergistic biological effects, although in such cases direct information as to the levels of particular pollutants would not be provided.

6 Impact of Sulphur Dioxide on the British Lichen Flora

6.1 Historical basis

The backcloth to studies on changes in the British lichen flora is (1) a literature, extending from the early decades of the seventeenth century, of over 2000 items and (2) over 200 000 specimens preserved mainly in museums. Study has not been equally intense in all parts of Britain throughout this long period. The most active phase of British lichenology prior to the formation of the British Lichen Society in 1958 were the Regency period and that of the Victorian naturalists (1850–90) and it is from the latter that the bulk of our data on past distributions derives. The British Lichen Society's Distribution Maps Scheme has stimulated field-work and co-ordinated information derived from this over the last decade. Most of England and Wales can now be considered fairly well known with respect to its extant lichen flora, although a few areas of Britain remain inadequately surveyed; the most important of these are most of Ireland, southern Scotland, Aberdeenshire, Caithness, Cheshire, south Lancashire, Shropshire, Staffordshire, and parts of central Wales (the lack of recent records from these areas must be considered when interpreting species distribution maps).

6.2 Declining species

There are many lichens in an active state of decline, at least locally, in Britain as over much of the Northern European plain. Some which are declining as a result of air pollution effects (in a few instances affected also by other factors; see Chapter 4) are listed in Table 7. Over very large areas of central England, from the London basin to Lancaster in the north-west and Newcastle upon Tyne in the north-east (see Fig. 5–3), nearly all corticolous and many siliceous rock species have completely disappeared in areas where they were formerly widespread. Different species are affected to varying extents by sulphur dioxide pollution (Table 5). *Usnea* species, formerly widespread and very common throughout lowland Britain (Fig. 6–1d), for example, have declined particularly dramatically; they were so common in counties such as Bedfordshire, Leicestershire, Nottinghamshire and Staffordshire during the last century that botanists listed them simply as 'common', 'general' or 'frequent', but where they do occur in the Midland counties today, specimens are often minute and

Table 7 Some species of corticolous lichens now very rare or absent in areas of England which now have mean winter sulphur dioxide levels over about 65 μg/m^3

Alectoria fuscescens	Lecidella elaeochroma	Peltigera horizontalis
Anaptychia ciliaris	Lepraria candelaris	Pertusaria spp.
Arthonia radiata	Lobaria spp.	Phaeographis dendritica
Arthopyrenia biformis	Nephroma laevigatum	Physcia aipolia
Bacidia rubella	Normandina pulchella	Physconia pulverulenta
Caloplaca cerina	Opegrapha atra	Pyrenula nitida
Candelaria concolor	Pachyphiale cornea	Ramalina calicaris
Enterographa crassa	Parmelia acetabulum	R. fastigiata
Graphis elegans	P. caperata	R. fraxinea
G. scripta	P. perlata	Sticta spp.
Gyalecta spp.	P. revoluta	Teloschistes flavicans
Lecanactis premnea	Parmeliopsis aleurites	Usnea spp.

on single trees in particularly sheltered sites. Some lichens, such as *Lobaria pulmonaria* and *Teloschistes flavicans*, started to decline at such an early date due to pollution and other human factors that the number of localized records is more sparse than for *Usnea*. While it is important to stress factors other than air pollution in cases of declines, these cannot explain losses from otherwise unaffected habitats. Although declines on a national scale have necessarily to be viewed as long-term, fluctuations almost from year to year occur in such species as *Hypogymnia physodes* close to their highest tolerated sulphur dioxide levels.

Although some forty lichens which have not been found in Britain since 1900 must be regarded as extinct, most of these were close to their geographical limits and their loss seems attributable to factors other than air pollution, such as small climatic fluctuations, habitat destruction and overcollecting.

6.3 Increasing species

The increase of sulphur dioxide tolerant species able to exploit man-made saxicolous substrates has been spectacular (Section 4.4; Fig. 6–2a, c), but in addition some corticolous lichens have also increased dramatically. *Lecanora conizaeoides*, a species with an efficient sulphur dioxide avoidance mechanism (Section 2.5), was not collected anywhere in the world prior to the middle of last century. By the latter decades of that century it was becoming widespread and it is now extremely common and abundant on trees and wood in most of Britain (Fig. 6–2b) and the north-west European plain where mean winter sulphur dioxide levels are in the range 55–150 μg/m^3; at levels above this it becomes rarer and at levels below 30 μg/m^3 of this pollutant it is also very rare. In addition to its tolerance, the elimination of competitors favours its abundance and in

§ 6.3 INCREASING SPECIES 39

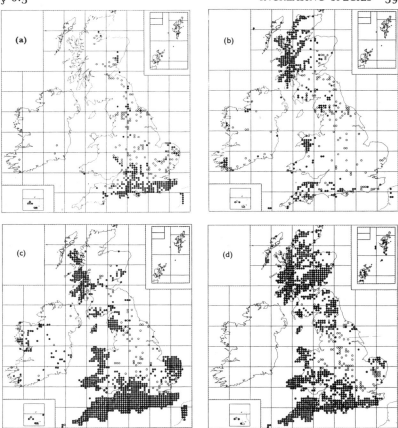

Fig. 6–1 Declining species. (a) *Anaptychia ciliaris*; (b) *Lobaria pulmonaria*; (c) *Parmelia caperata*; (d) *Usnea* spp. (omitting Ireland). o = pre-1960; ● = post-1960. Based on data from the British Lichen Society's Distribution Maps Scheme.

relatively unpolluted areas it usually occurs on man-made substrates (e.g. fence-posts) and twigs, failing to enter communities already established on mature tree trunks. The elimination of competitors may also have encouraged expansion in a few sulphur dioxide-tolerant species found on man-made substrates in urban areas; *Lecanora muralis*, for example, remains uncommon in less polluted rural areas of southern and western Britain (Fig. 6–2c).

Acidification of tree bark by sulphur dioxide pollution has also led to the increase of some species characteristic of acidic barks in unpolluted areas. *Parmeliopsis ambigua*, for example, formerly almost entirely

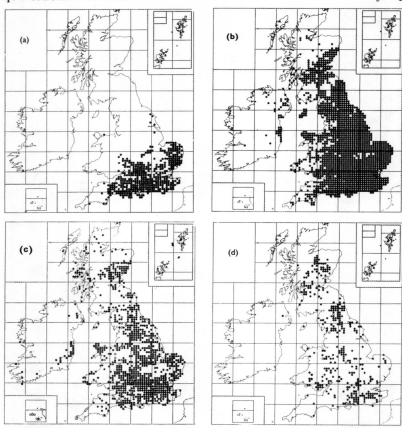

Fig. 6-2 Increasing species. (a) *Caloplaca teicholyta*; (b) *Lecanora conizaeoides*; (c) *L. muralis*; (d) *Parmeliopsis ambigua*. o=pre-1960; ●=post-1960. Based on data from the British Lichen Society's Distribution Maps Scheme.

restricted to the acidic bark of coniferous trees, is now frequent in many parts of lowland Britain subject to mean winter sulphur dioxide levels in the range 55–65 $\mu g/m^3$ (Fig. 6-2d) where it occurs on a wide range of deciduous trees on which it was formerly unknown in Britain. This effect may also contribute to the abundance of some other species in moderately polluted areas (e.g. *Hypogymnia physodes*, *Platismatia glauca*, *Pseudevernia furfuracea*).

6.4 Species numbers

The number of corticolous species in a site is determined by woodland management factors (Section 4.2) as well as by air pollution. A

§ 6.4 SPECIES NUMBERS 41

comparison of sites having mature trees in lowland Britain, particularly ancient parkland, shows quite a good correlation with mean winter sulphur dioxide levels; species numbers decline as all major polluted areas are approached (Fig. 6–3). Historical data show that many such sites in areas now polluted had many more species in the past than they do today. Epping Forest (Essex) is so well documented that its decline from zone 9 or 10 in 1784–96 to 3–4 in 1974 can be followed; at least 120 corticolous and lignicolous species were present up to the middle of last century but in 1970–74 only 38 such species were to be found in the whole forest (only 18 in the part nearest London). Declines are also documented over much shorter periods; Bookham Commons (Leatherhead, Surrey), for example, lost 12 corticolous species between 1953–56 and 1969–73, a decline of 25%. In whole counties which were well studied during the last century and which are now affected, very many species have been

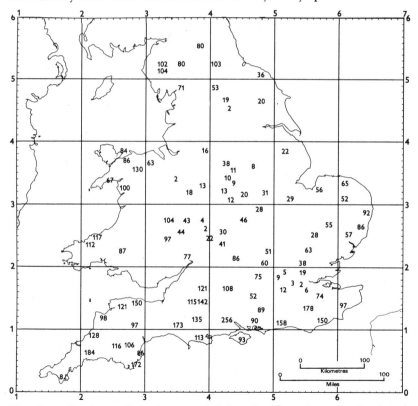

Fig. 6–3 Numbers of species recorded in selected ancient woodland and parkland sites in England and Wales since 1967.

42 OTHER EFFECTS § 6.5

lost completely; 154 species (47% of its total lichen flora) have not been seen in Leicestershire in recent years. Declines of this order are not seen in areas remaining largely unaffected such as Devonshire and Sussex which show declines of 15% and 18% respectively, almost entirely due to other factors. In a suburb of Torquay, out of the 207 species recorded, only 18 were not seen in 1973 (mean winter sulphur dioxide level about 28 $\mu g/m^3$).

6.5 Other effects

Reductions in thallus size and production of ascocarps (Section 2.3) are seen in many species; production of ascocarps in some species even when of normal sizes proves of value in zone scales for estimation of sulphur dioxide levels (e.g. *Anaptychia ciliaris* in Table 5). Changes in fertility can also be plotted on maps, and even in relatively unpolluted areas ascocarps are now much rarer in many species which are still widely distributed. *Evernia prunastri*, for example, used to produce apothecia as close to London as Virginia Water but now its closest fertile station to London is in Dorset. Reduction in size, also seen in many species (e.g. Fig. 6–4), may result from die-back of parts of thalli, reduced growth rates and other factors (e.g. fluctuating populations, reduction in photosynthetic rates).

In addition to relict lichens resulting from changes in woodland

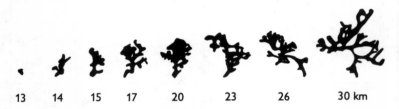

Fig. 6–4 Specimens of *Evernia prunastri* (silhouettes, × ½) collected from ash trees at different distances from the centre of Newcastle upon Tyne (based on data of Dr O. L. Gilbert).

(Section 4.2) and heathland (Section 4.3) management, air pollution gives rise to relict populations. These, in contrast to actively declining species, are able to persist where established while the current environment remains unchanged, but are unable to colonize new substrates presented to them in the now hostile environment. In the Midland counties, in areas with moderate sulphur dioxide pollution, it is not uncommon to find ancient ash trees or stumps bearing species absent on younger nearby trees. Similarly, on tombstones in polluted areas some species (e.g. *Caloplaca heppiana*) are found to be restricted to those erected prior to a particular date (e.g. 1900).

7 Other Considerations

7.1 Air pollution trends

Data from the some 1200 recording gauges in the British Isles show that mean *urban* sulphur dioxide levels have fallen since the Clean Air Act of 1956 (Fig. 7–1). Total emissions remain at a very high level and have risen over this period, largely due to increased emissions from electricity

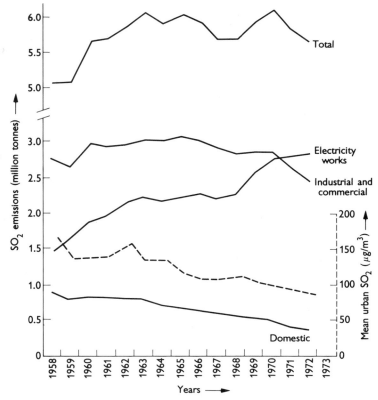

Fig. 7–1 Sulphur dioxide emissions and sources compared with mean urban sulphur dioxide levels in the British Isles 1958–73 (based on data provided by the Warren Spring Laboratory).

works, which more than compensate for reductions achieved in those from other sources. The siting of new power stations far from urban areas and the use of very high chimneys mean that the bulk of their sulphur dioxide is dispersed over wide areas, with only marginal effects on urban gauges. The marked declines in mean levels found in urban areas do not seem to have occurred in gauges sited in rural areas. A great deal still has to be achieved before our urban areas can be considered clean; a maximum monthly average sulphur dioxide level of 135 $\mu g/m^3$ is regarded in Sweden as one not to be exceeded, yet in the winter months many British towns have levels very much higher than this. Natural background sulphur dioxide levels are probably in the range 0.28–2.8 $\mu g/m^3$.

7.2 Effects on other plants

Sensitivity in flowering plants varies in different species; sulphur dioxide acts on chlorophyll, as it does in lichen algae, leading to chlorosis and necrosis (Section 2.2). Most higher plants are much more tolerant than lichens, but coniferous trees appear to be particularly sensitive; adverse effects are reported at mean sulphur dioxide levels of only 56 $\mu g/m^3$ in Sweden, and where these exceed 196–224 $\mu g/m^3$ in the Ruhr they fail to grow. Effects on native flowering plants in Britain remain almost completely unknown. Both fluorides and components of photochemical smog cause severe damage to crop plants. Secondary effects can also arise from sulphurous rain leading to an increased leaching of the soil and the inhibition of nitrogen-fixing bacteria.

Some mosses and liverworts (bryophytes) are probably as sensitive to sulphur dioxide as the most susceptible lichens (e.g. *Antitrichia curtipendula*) while others are more tolerant and penetrate far into urban areas (e.g. *Orthodontium lineare, Tortula muralis*). These can be used in biological scales for the estimation of sulphur dioxide levels to complement the lichen data. Effects on algae are almost unknown, although at least one is encouraged by smoke (Section 3.1). Some leaf-inhabiting fungi are adversely affected by sulphur dioxide and may be eliminated altogether; some mildews, many rusts, *Rhytisma acerinum* (Tar-spot of sycamore) and *Diplocarpon rosae* (Black-spot of roses) are limited by mean levels of about 100 $\mu g/m^3$).

7.3 Effects on man and his materials

High concentrations of gaseous sulphur dioxide (2700–13 500 $\mu g/m^3$) cause irritation of air passages, but small droplets or particles containing sulphurous compounds are more harmful; these may have contributed to the increased mortality rate of some 4000 people during the London 'smog' of 1952 when exceptionally high smoke (4500 $\mu g/m^3$) and sulphur dioxide (35 100 $\mu g/m^3$) levels occurred together.

Corrosion rates of metals are increased by sulphur dioxide; these are 2–5 times faster (depending on the metal and coating) in large towns in Sweden (mean sulphur dioxide levels about 100 $\mu g/m^3$) than in rural areas (about 8 $\mu g/m^3$). Calcium carbonate of limestones and calcareous sandstones is slowly converted to calcium sulphate which renders stones more susceptible to deterioration; damage to historically important buildings, particularly to carved stone, is particularly regrettable.

7.4 Effects on other organisms

About 27 European moths are dependent on lichens for food or camouflage at different stages of their life cycles. Where pollutants have eliminated lichens some moths have been able to produce dark-coloured (melanic) races much less conspicuous to predators on lichen-free trees than their paler lichen-camouflaged counterparts (e.g. *Biston betularia*). Although entomologists have expressed concern over such indirect effects little precise information is available; the invertebrate fauna of trees does, however, become reduced in diversity (number of species) as polluted areas are approached and the lichen flora shows less variety. Lichens also provide food and shelter for many less conspicuous invertebrates (including species of Acari, Coleoptera, small Lepidoptera, Mollusca, Psocoptera and Rotifera) and these will obviously suffer as the lichens disappear.

7.5 Conservation

The complexities of ecosystems in terms of the interdependence of the component species are only just becoming appreciated, and it is almost impossible to forcast how a change in the population of one organism will affect another. Consequently, modern conservation practice aims at preserving intact ecosystems; by ensuring the survival of one species the survival of others directly or indirectly dependent on them will be facilitated. This thesis is applicable to lichens (Section 7.4), which are also important in conservation policies because they are the most sensitive organisms to sulphur dioxide known. By ensuring the survival of the most susceptible lichens one can be reasonably confident that other organisms are not being affected by this pollutant. It is for this reason that Professor K. Mellanby has stressed that we should work towards 'air fit for lichens'. Sulphur dioxide pollution is now so widespread that its total elimination would involve astronomically prohibitive expenditure. Steps should, however, be taken to ensure that conditions do not deteriorate further in areas currently little affected (Fig. 5–3).

Appendix 1 : Lichen Identification

A1.1 Collection and preservation

Lichens are one of the easiest groups of plants to collect and preserve. Species are removed, with part of the substrate, with the aid of a sheath-knife (for those on trees this is done by cutting only into the dead outer layers of bark to minimize damage to the tree), or a geological hammer of about 1–2 lb. (0.90 kgs) in weight and a small cold-steel masonry chisel of about $\frac{1}{4}$–$\frac{1}{2}$ in (6 to 12 mm) in width (for those on rock). Most can be seen with the naked eye but a hand-lens ($\times$10) is necessary to avoid overlooking tiny crusts and is also valuable in field identifications. Specimens should be representative and whole specimens rather than fragments taken to ensure that they will be adequate for naming and reference. Once collected, specimens are placed in tins, paper or polythene bags, together with details of the locality, substrate and date: fragile specimens are best first wrapped in tissue paper. Shortly afterwards the specimens should be thoroughly dried (e.g. by placing on a radiator overnight); slight colour changes occur in some species, but most appear as they do under dry conditions in the field. Once dried, specimens will keep indefinitely, provided they are stored in a dry place. Each should be placed in a separate paper packet or envelope bearing full details of the locality, etc.; packets themselves can be filed alphabetically in shoe-boxes or filing cabinets (the more fragile being glued, with some parts upside down so that both surfaces can be seen, on index-cards and covered with tissue paper). For further information on these aspects see Duncan's book (Section A1.3).

A1.2 Conservation

Collectors must always have the word 'conservation' in their minds. Care must be taken not to eliminate a species from a site, and the greater part of a colony should be left untouched. This is particularly important in moderately polluted areas where species may be relics (Section 6.5) and may be eliminated locally by careless collectors. For this reason it is best to start to learn to identify lichens and build up a reference collection in areas where species are still very common before starting to survey polluted areas; if this is done few specimens need be collected during such surveys.

Permission to collect from privately owned trees and walls must always be obtained so as not to prejudice owners against other naturalists.

Collecting from tombstones is not permissible but notes, chemical tests and even slides can be prepared *in situ*.

A1.3 Naming

Most lichens are no more difficult to identify than the majority of flowering plants—indeed with practice most can be reliably named in the field. Initially it will be necessary to learn a few terms (Sections 1.4–5, A1.4), the names of some chemical reagents and how to use a microscope. Mistakes will be made at first, but as a reference collection is built up and field experience grows these will become fewer. The best way to learn to recognize lichens is to attend a field course or trip led by experts.*

Newcomers to lichens often start by learning about the larger foliose and fruticose species, progressing to the crustose ones later, although many of these are easily identified. When examining specimens, the size and form of the thallus, its colour, any markings or special structures (e.g. isidia, soralia) and their form and position, and the type of any ascocarps present are the first things to note; then chemical reactions (Section A1.4). Once key characters are known most species can be identified with the aid of a hand-lens (×10). The following books are useful to beginners identifying British lichens.

ALVIN, K. A. and KERSHAW, K. A. (1963). *The Observer's Book of Lichens*. 126 pp. Warne, London. [Out of print but in many libraries.]

BRIGHTMAN, F. H. and NICHOLSON, B. E. (1966). *The Oxford Book of Flowerless Plants*. 208 pp. Oxford University Press, London. [£3.95; includes coloured illustrations.]

DAHL, E. and KROG, H. (1973). *Macrolichens of Denmark, Finland, Norway and Sweden*. 185 pp. Scandinavian University Books, Oslo, etc. [£3.95; foliose and fruticose species.]**

DUNCAN, U. K. (1970). *Introduction to British Lichens*. 366 pp. Buncle, Arbroath. [£4; the standard flora essential to all serious students.]**

---

* Week-long courses at various centres in Britain are run by the Field Studies Council (9 Devereux Court, Strand, London WC2R 3JR); details of these, other courses and field meetings are included in the *Bulletin* of the British Lichen Society (*Secretary*: J. R. Laundon, Department of Botany, British Museum (National History), Cromwell Road, London SW7 5BD). The Society has a panel of referees who will assist in naming specimens. It also has a collection from which named specimens can be borrowed, and issues a journal, *The Lichenologist*, in addition to the *Bulletin*. Subscriptions (for 1976): ordinary members £5 p.a. (including the journal); junior associate members £1 (including the *Bulletin* but not the journal).

** The sole British agent for these is the Richmond Publishing Co., Orchard Road, Richmond, Surrey, TW9 4PD.

A1.4 Some species used in air pollution surveys

It is not necessary to have a comprehensive knowledge of British lichens to carry out surveys in polluted areas. Some studies require that just a few species are properly known while other exercises can be carried out without naming any to species level (See Appendix 2). A few notes on the more important features of some species used in air pollution survey work are given here which should enable them to be recognized quickly in the field. Most lichens included are primarily corticolous but some occurring on rocks are also mentioned. The zones (Table 5; Fig. 5–3) in which a species *first* becomes frequent on *trees* are indicated in brackets after the name, e.g. *L. incana* (3); the British distribution of some species treated here is shown in Figs 6–1 and 6–2.

Most terms used here have been defined above (Section 1.4–5); some others are explained when first mentioned. Notes on spores are included where helpful (examine by squashing a small slice of an ascocarp in K, see below). Remember that close to their innermost stations to pollution sources specimens may be small and lack ascocarps, otherwise usually present (Section 2.3). The following chemical reagents will be found useful in identification:

K = 15% aqueous solution of potassium hydroxide. Positive reactions: yellow, yellow changing to red, red, brownish or crimson.

C = commercial liquid bleach (e.g. 'Parazone', 'Domestos'); observe immediately. Positive reactions: red, rose, orange or green.

KC = a drop of K followed after blotting by one of C. Positive reactions: red, rose or purple.

PD = *p*-phenylenediamine; a toxic compound best used as a solution (1 g PD: 10 g sodium sulphite: 100 ml water: about 10 drops liquid detergent; this is stable for about 6 months). Allow 3–5 min for colours to develop. Avoid contact with paper, packets, books, etc., which will stain. Positive reactions: yellow, yellow changing to red, or orange.

These tests are made by removing a small fragment of the lichen, placing it on filter paper, adding a drop of reagent, and noting colour changes (×10 lens useful); tested fragments must be discarded. When performing tests on the medulla the cortex is first scraped away with a razor blade. To avoid accidents chemicals should *not* normally be carried in the field; when this is necessary (e.g. when examining tombstones) securely fastening plastic droppers are recommended (PD is rarely needed in such cases).

Leprose and crustose species

Lepraria species never have ascocarps: *L. incana* (3) forms greyish-green to blue-green powdery patches on shaded non-calcareous walls and trees (K+, PD–, i.e. positive to K and negative to PD); *L. candelaris* (5) forms a bright golden-yellow powder most commonly in bark crevices (K–, PD–). The emerald green alga *Pleurococcus viridis* (1) is often common, covering whole trees in soot-polluted areas (Section 3.1) and may

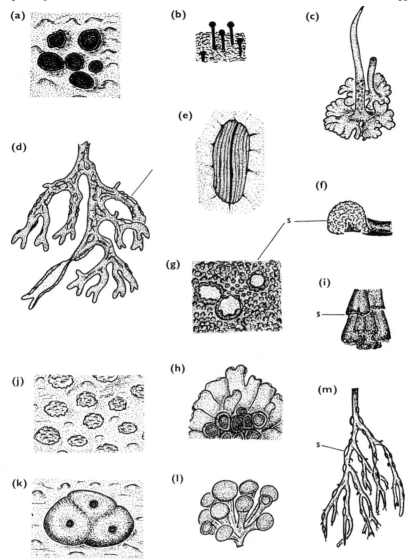

Fig. A1–1 (a) *Buellia punctata* (×15); (b) *Calicium viride* (×3); (c) *Cladonia coniocraea* (×2); (d) *Evernia prunastri* (×1); (e) *Graphis elegans* (×10); (f) *Hypogymnia physodes*, sorediate lobe-end (×10); (g) *Lecanora conizaeoides* (×15); (h) *L. muralis*, angular apothecia (×7); (i) *Lecidea scalaris* (×10); (j) *Pertusaria amara* (×7); (k) *P. pertusa*, wart with apothecia (×15); (l) *Ramalina fastigiata* (×1); (m) *R. farinacea* (×1). s = soralia.

50 SOME SPECIES USED IN AIR POLLUTION SURVEYS § A1.4

Fig. A1–2 (a) *Lecanora dispersa* (×3½); (b) *Xanthoria parietina* (×3); (c) *Lecanora muralis* (×5.5); (d) *Physcia adscendens* (×2); (e) *Parmelia saxatilis*, lobe ends showing white net (×3). Photographs by F. S. Dobson.

§ A1.4   SOME SPECIES USED IN AIR POLLUTION SURVEYS   51

Fig. A1–3 (a) *Hypogymnia physodes*, young plant (×2); (b) *Lecanora chlarotera* (×6½); (c) *Parmelia caperata* (×1½); (d) *Usnea articulata* (×¼). Photographs by F. S. Dobson.

sometimes also look leprose. Some crustose lichens are almost entirely dissolved into soredia so may easily be mistaken for leprose species, e.g. *Lecidea lucida*, forming bright yellow stains on brickwork and *Lecanora expallens*, forming neat C+ orange (i.e. C positive reaction) yellow-green patches on trees (apothecia yellow-green to flesh coloured). Black pin-heads on a bright green thallus in bark crevices characterize *Calicium viride* (5; Fig. A1–1b).

Lecanora species, like those discussed in the remainder of this section, are crustose. Their apothecia have margins containing algal cells and look like jam tarts. *L. conizaeoides* (2; very common 2–5, rarer 6–10 (Fig. A1–1g)) often forms grey-green swards covering whole trees (3–4); its rather coarse thallus is often almost completely covered with soredia coming off as a powdery deposit on fingers and clothing (K–, C–, PD+ bright red; apothecia common). *L. dispersa* (Fig. A1–2a), lacking soredia, is very variable with a thallus reduced to a black stain in soot-polluted areas but with a white thallus in unpolluted ones; it is always fertile with apothecia 0.5–1 mm diameter which have yellow-brown to brown discs and white margins (very common on concrete and other base-rich substrates). *L. chlarotera* (5; Fig. A1–3b, only occuring on trees), also lacks soredia and has apothecia with white margins but here the discs are red-brown and the thallus is greyish.

Candelariella vitellina (4) has a thin to thick mustard yellow K– thallus and apothecia with small yellowish-green discs (K–; spores colourless, simple, 12–32 per ascus) and occurs on trees and siliceous rocks. *Caloplaca citrina* is orange-yellow, sorediate and K+ crimson. Similar K– discs on a blackish stain on concrete are *C. aurella*.

Buellia punctata (3; Fig. A1–1a) on basic barks has small (0.5 mm diameter) black apothecia with convex discs and a disappearing proper margin (i.e. lacking algae) on a greyish thallus (K–, C–, PD–; spores brown, one septate).

Pertusaria pertusa (5; Fig. A1–1k) forms almost circular grey patches with somewhat zoned and white margins on trees and has the apothecia immersed 2–3 together in rounded warts. *P. amara* (5; Fig. A1–1j) lacks apothecia, but has white disc-like soralia 0.5–1.5 mm diameter and is easily identified by the bitter taste left on the tongue if soredia are applied with a moistened finger (KC+ violet; *P. albescens* (6) does not taste and is KC–).

Graphis elegans (6; Fig. A1–1e) is the only British species with elongate black lirellae with lip-like margins each of which has 2–5 longitudinal grooves (spores colourless, 10–12 septate).

Squamulose species

Overlapping minute pale brownish rather convex C+ reddish squamules with sorediate margins characterize *Lecidea scalaris* (4; Fig. A1–1i) which occurs on palings, trees, siliceous rocks and bricks.

Rosette-like (placodioid) thalli are produced by a number of distinctive species used in air pollution studies:

Buellia canescens (3) forms thick circular white to greyish hoary (K+ yellow) patches 2–4 cm across which tend to break away in the centre, the lobes are most distinct at the margins in this species which is common on basic rocks and barks.

Some saxicolous *Caloplaca* species form rather similar rosettes which are K+ crimson and yellow or orange in colour. *C. aurantia* has egg-yolk yellow flattened marginal lobes whilst *C. heppiana* has deep orange markedly convex marginal lobes. *C. teicholyta*, however, is grey and rather similar to *B. canescens* but is flatter, not breaking away centrally, and K−. *Candelariella medians* forms yellow rosettes not unlike those of *C. heppiana* but is paler in colour and K−.

Lecanora muralis (Fig. A1–1h, Fig. A1–2c), common on man-made substrates in polluted areas, forms yellowish-brown to greenish-brown circles often 2–10 cm across; apothecia are usually present, have yellowish brown to reddish brown discs, white margins, and are often angularly compressed.

Foliose species

Hypogymnia (*Parmelia*) *physodes* (4; Fig. A1–1f and A1–3a) has smooth, narrow (0.2–0.4 cm wide) inflated and convex lobes (pale grey above the black below), lacks rhizinae, and has soredia on the lower side of often raised and upturned lobe ends (medulla C−, KC+ red, PD+ orange).

Lobaria includes some of the largest foliose lichens; all species have a felt-like undersurface lacking neat circular pits (cyphellae) and without rhizinae. *L. pulmonaria* (9), 'Lungwort', has bright-green (wet) to brownish (dry) lobes (often 8–15 cm long) attached at one end and spreading over large areas; the lobes are deeply furrowed between distinct ridges which bear isidia or soredia (medulla K+ yellow changing to red).

Parmelia (spores, when present, colourless, non-septate), the largest genus of foliose lichens, includes species with flattened lobes attached to the substrate by rhizinae; most form rosettes. *P. saxatilis* (4) and *P. sulcata* (4) both have a distinct network of coarse white ridges on their grey lobes (Fig. A1–2c), and a medulla reacting K+ yellow changing to red; the former has isidia on the ridges while the latter bears soralia instead. *P. perlata* (8), in contrast, has quite smooth pearly-grey lobes which are rounded and undulate and bear conspicuous creamy soredia on their margins and ends (medulla K+ yellowish, C−). The common *P. caperata* (6; Fig. A1–3c) is easily recognized by its large (often 5–10 cm diameter) adpressed yellow-green rosettes with rounded, regularly indented lobes smooth at the margins but wrinkled and sorediate centrally (medulla K+ yellow). All the above have lobes 0.5–1 cm wide, but *P. glabratula* (5) has tightly adpressed dark brown to almost black, smooth, shiny, narrow (0.2–0.4 cm) lobes supporting dense masses of rod-like isidia (medulla

C+ red); *P. exasperatula* (6), mainly a northern species, is rather similar, but here the isidia are hollow and club-shaped (medulla C−).

Parmeliopsis ambigua (5) is similar in colour to *Parmelia caperata*, but forms much smaller rosettes (1–3 cm diameter) and has short, narrow (0.1 cm wide), deeply incised rather square-ended lobes and bears large convex powdery masses of soredia.

Physcia species (spores, when present, brown, 1-septate) are generally narrower and often thicker than those of *Parmelia* and prefer basic barks and rocks. Small pale grey tufts of rather convex lobes with stout bristles arising from their margins occur in *P. adscendens* (4; Fig. A1–2d) which bears soredia under helmet-shaped lobe-ends; *P. tenella* (5) is similar but here the soredia occur below upturned (not helmet-shaped) lobes. Small adpressed grey-brownish lobes (bright green when wet) with fine marginal hairs and greenish to black convex soralia on the lobe surfaces characterize *P. orbicularis* (5). *P. aipolia* (7) lacks hairs on the lobe ends, has no soredia, commonly has apothecia with blue-black discs and white margins, and forms neat rosettes of adpressed, radiating, grey lobes with tiny white dots (lens) on their surface.

Platismatia (*Cetraria*) *glauca* (5) has a rather thin and papery waxy-grey to brownish tinged thallus with ascending incised crisp lobes often somewhat cracked on the surface and lacking rhizinae below; the lobes are fringed with isidia or soredia, and the medullary reactions (K−, C−, PD−) easily distinguish it from *Parmelia*.

Peltigera species, unlike those mentioned above, usually occur on the ground. The lower surfaces are felt-like and often have distinct veins from which stout rhizinae arise. *P. canina* has a surface which is grey (dry) and rather downy (lens) with large convex swollen (bullate) patches and down-turned margins.

Xanthoria species are orange-red and K+ crimson. *X. parietina* (4; Fig. A1–2b) forms rosettes often 2–10 cm diameter and is nearly always fertile. *X. candelaria* (5), in contrast forms small tufts of erect lobes (about 1 mm wide) with yellow soredia on their margins: the superficially rather similar *Candelaria concolor* (7) is distinct in being K−.

Fruticose species

A few species which appear fruticose are not strictly so anatomically, having different upper and lower surfaces. Three species of interest here fall in this category. *Evernia prunastri* (5; Fig. A1–1d) with strap-like, often contorted, flat, forked lobes to 6 cm long which are greenish to yellowish-grey and net-marked above, but powdery white below (unlike *Ramalina* species). *Pseudevernia furfuracea* (5), a species of acidic substrates, has 3–4 mm wide strap-like lobes pale grey (not yellowish) above and with older lobes black underneath; coral-like isidia are present and the medulla is C+ red or C−; *E. prunastri* is C−. *Anaptychia ciliaris* (7), frequent on basic barks in the east, is also strap-like but here the lobes are downy, pale grey, fringed with stout bristles and whitish below (apothecia in zones 8–10 with frosted brown-black discs).

Cladonia species fall in a special category also. Most of those found on trees have small greenish often incised squamules which are white below and 1–3 mm across, and from these horn-, antler- or goblet-like secondary structures (*podetia*) arise vertically (often to 0.5–2 cm tall). Many species are involved (e.g. Fig. A1–1c; see Duncan; Section A1.3).

Alectoria fuscescens (6), occurring on acidic substrates mainly in upland areas, is truly fruticose, forming smoky to dark brown much-branched entangled hair-like usually pendent plants often 3–10 cm long; this also has whitish-green soralia (K–, PD+ red). In *Alectoria* the centre of the branches is hollow, unlike *Usnea* which has a solid central core (seen by scraping away the surface with a finger-nail; Fig. 1–1a). *Usnea* species, 'Beard lichens', are usually yellowish green or greenish grey and may be pendent, trailing, or tufted depending on the species. *U. subfloridana* (7) forms shrubby tufts 1–8 cm tall, has a stout blackened base and tufts of spiky isidia often becoming broken off to leave soralia-like patches (PD+ orange). *U. florida* (9) lacks isidia and usually has massive (1–2 cm diameter) fawn apothecia, but is tufted like *U. subfloridana*. *U. ceratina* (8) has stout stems with wart-like tubercles and has a trailing to pendent, often scrawny habit (to 40 cm long; PD–). *U. articulata* (10), still common locally in the south-west, is the largest British *Usnea* (to 1 m) readily recognized by the trailing to pendent habit and inflated sections to the stems, making it appear like strings of grey sausages (Fig. A1–3d).

Ramalina species have greenish grey or yellowish green thalli which are often flattened but with similar upper and lower surfaces and no central axis. *R. farinacea* (5; Fig. A1–1 m) is the most pollution-tolerant of the genus and has narrow pendent lobes 1–3 mm wide with distinct K– saucer-shaped soralia along their margins. *R. fraxinea* (9) has strap-like, often broad, pendent branches with flattened lobes and a markedly wrinkled surface; this lacks soralia and usually has fawn apothecia on the lobe margins. *R. fastigiata* (8; Fig. A1–1l) forms rather dense erect tufts 1–3 cm tall composed of rather inflated lobes lacking soralia and bearing terminal apothecia.

Appendix 2: Practical Exercises

A2.1 Introduction

This Appendix outlines some practical studies which can be carried out by schoolchildren, or, at a more sophisticated level, by undergraduate or college students and adult amateurs. As parts of courses concerned with the environment such studies will serve both to illustrate the effects of man on living organisms and to demonstrate the value of biological monitoring in pollution surveys. Furthermore, they may be of scientific value, adding to our knowledge of pollution patterns in the areas investigated. Most of the exercises mentioned here require little or no previous knowledge of lichens; indeed in some there is no need to identify any of the species involved. As evidence of the small background experience required the ACE survey by schoolchildren should be stressed (Section A2.4). Undergraduates wishing to use more complex scales (e.g. Table 5) will find the notes on the identification of species (Section A1.4) helpful. In the course of all studies mentioned here remember *conservation* (Section A1.2).

The sharpness of field results obtained will tend to be proportional to the steepness of pollution gradients present. In the case of all studies mentioned here it will be instructive to relate the lichen evidence to that derived from sulphur dioxide recording gauges in the study area, when these are present (for the sites and results from gauges contact either your local authority or the Warren Spring Laboratory, Gunnels Wood Road, Stevenage, Herts., SG1 2BX); mean annual or mean winter values averaged over the last 3–4 years should be used.

A2.2 Species numbers

1 a Count the number of species on one selected substrate which seem different and repeat in other sites in a line or belt transect along a suspected pollution gradient; there is no need to name the lichens involved. Plot the total numbers on a graph with distance as the horizontal axis; suitable substrates include (i) limestone tombstones over 100 years old, (ii) old asbestos-cement roofs, and (iii) the trunks of one kind of tree (e.g. ash, oak, sycamore). **b** If enough sites are available over a large area plot a map using contours of maximum lichen counts. In **a** and **b** remember that numbers derived from each substrate must be treated separately; if several have been looked at compare the differences (how and why do they differ?). c Using tombstones compare the numbers on those over 100 years old with those erected since 1900 (Section 6.5).

2 Record the height to which a particular species extends up the trunks of one kind of tree (e.g. ash, oak) along a transect. Several species can be used but the results from each must be kept separate.

3 As many biologists find the interaction of organisms fascinating, and to show how a declining lichen flora affects quite different organisms, count the number of different kinds of insects you can find associated with lichens along your transect (Section 7.4).

4 For more striking contrasts, compare the numbers of species you can find on the *same* substrate in two quite different areas (e.g. in your local city or suburbs and while on a holiday or field course in less polluted areas of western Britain).

A2.3 Distribution of growth forms

Record along a suitable transect along a suspected pollution gradient (e.g. a town centre out into the surrounding country) or at scattered sites over this area (if you wish to show the results as a map rather than a graph) one or more of the following (keeping the results from each separate):

1 a Green-grey crustose lichens on trees; b grey or white foliose lichens on trees; or c shrubby (fruticose) lichens on trees.

2 Orange foliose lichens (*Xanthoria*) on a asbestos-cement roofs; b tiled roofs and brickwork; or c tree trunks.

3 Rosettes of the yellow-grey squamulose lobed *Lecanora muralis* on a asbestos-cement roofs; or b brickwork or tiled roofs (see Table 6).

4 If in a mainly rural area compare a the distribution on a transect from a town or large village into the country of grey or white foliose lichens on (i) elm or sycamore and (ii) lime or oak (do they differ; if so, why?). b Assess the percentage cover (by eye or with the aid of tracing paper) of particular growth forms. c To what height up the tree does each growth-form ascend in different sites? d Attempt to identify the genera (or species) concerned. Plot separate graphs for the data from a–c; what differences occur between the two groups of trees in a and why? (See Sections 1.8, 2.4.) e Repeat with ash trees if possible (are these in groups (i), (ii) or a different group?). f If you can find trees which are shaded or in deep valleys are there differences in the percentage covers of growth forms and the species present? g Is there any difference in the percentage of mosses in f?

A2.4 ACE zone scale

A simplified zone scale (Table 8) was used in the Advisory Centre for Education's (ACE) air pollution survey carried out by schoolchildren in 1972 with the aid of their *Clean Air Research Pack* (97p each, post free, from

Table 8 The ACE zone scale (adapted from GILBERT, O. L., *Environ. Pollut.*, 6, 175–180, 1974)

Zone	Lichens and mosses	Mean winter SO$_2$ (μg/m^3)
0	*Pleurococcus* growing on sycamore	Over 170
1	*Lecanora conizaeoides* on trees and acid stone	About 150–160
2	*Xanthoria parietina* appears on concrete, asbestos and limestone	About 125
3	*Parmelia [saxatilis/sulcata]* appears on acid stone and *Grimmia pulvinata* occurs on limestone or near mortar	About 100
4	Grey leafy [foliose] species [e.g. *Hypogymnia physodes*] start to appear on trees	About 70
5	Shrubby [fruticose] lichens [e.g. *Evernia prunastri*] start to appear on trees	About 40–60
6	*Usnea* becomes abundant	About 35

ACE, 32 Trumpington Street, Cambridge CB2 1QY) which includes an identification chart. This was used very successfully by twelve to fourteen-year-olds to produce a map for the British Isles comparing favourably with those formed by more complex methods. Further information on this survey and its results are included in:

GILBERT, O. L.(1974). Air pollution survey by school children. *Environ. Pollut.* 6, 175–180. [Pp. 159–80 available separately at 20p; includes details of the ACE water pollution survey.]

JACKSON, B. and YOUNG, G. (1973). Watching what we breathe. *The Sunday Times Magazine*, 28th January 1973, 40–46.

MABEY, R.(1974). *The Pollution Handbook. The ACE/Sunday Times Clean Air and Water Surveys.* 144 pp. Penguin Education, Penguin Books, Harmondsworth. [70p; includes coloured photographs of species used.]

RICHARDSON, D. H. S. (1975). *The Vanishing Lichens.* 231 pp. David & Charles, Newton Abbot, London and Vancouver. [£5.95.]

A2.5 Transplants

1 Collect specimens of a range of species of different sensitivities (see Table 5, Section A1.4) on small slivers of bark or twigs and glue with Araldite (CIBA Ltd) on to similar trees where they are absent (*conservation*, Section A1.2). If sufficient material can be collected (*conservation*, Section A1.2) place samples along a transect. Note colour changes, relative rates of death, etc. (Section 2.4) over several months at two-week intervals (time will vary according to the weather and pollution levels; why?).

2 Repeat with pieces of rock (not from walls or private property!) placing specimens at the same inclination and aspect. Compare changes

of lichens on limestone (if any) with those on siliceous rocks (e.g. sandstone, granite).

3 If in an area lacking suitable trees, take samples as in 1 and glue to about 20 × 20 cm plywood squares on posts at several hundred yard intervals on a transect; boards must be vertical and all face the same way (study as in 1, above).

A2.6 Detailed species mapping and listing

1 Record *all* occurrences of *one* certainly identified species on *one* type of substrate (e.g. *Parmelia caperata* on oak) along a suspected pollution gradient or over a large area. Plot the sites which had the species on a map, using a different symbol to indicate sites of suitable habitats where the species was not found. This study could be extended to cover other species either more or less sensitive than that used initially. The innermost limits of several can be shown on the same map (Fig. 5–1). If pollution gradients are steep this study will be particularly instructive.

2 Older or more experienced students and amateurs who have learnt a large proportion of the lichens in their areas may like to try to build up a list of those now present there. Initially it is easiest to study one substrate (e.g. trees) first. Data from such surveys should be submitted to the British Lichen Society's Distribution Maps Scheme.* At the same time try to build up a list of species found in the area in the past from published accounts and preserved specimens which had been collected long ago (your local museum should be able to help here). Do any species seem to have become extinct? Have any become rarer or more common? What factors are responsible (see also Chapter 4)? Studies of this type are particularly valuable in the Midlands and other areas well surveyed in the past, or where new pollution sources are about to open or have recently opened so that declines can be observed. Institutions such as schools, colleges, polytechnics, universities or local natural history societies may like to carry out surveys of the same area at five- or ten-year intervals to record precisely any changes that occur. Permanent quadrats (marked out squares or rectangles on trees or walls) can also be photographed at intervals. Voucher specimens should be collected (*conservation*, Section A1.2) and carefully preserved (e.g. in a local museum) so that future workers can check them (names and concepts of species change over the years).

*Dr. M. R. D. Seaward, School of Studies in Environmental Science, University of Bradford, Bradford BD7 1DP.

Further Reading

Chapters to which the following relate are indicated after the references. Further references dealing with identification and practical exercises are cited in the Appendices.

AHMADJIAN, V. and HALE, M. E., eds. (1974) ('1973'). *The Lichens*. 697 pp. Academic Press, New York and London. (Chs. 1–3, 5)

BROWN, D. H., HAWKSWORTH, D. L. and BAILEY, R. H., eds. (1976). *Lichenology: Progress and Problems*. (Systematics Association Special Volume No. 8.) 551 pp. Academic Press, London, New York and San Francisco. (Chs. 1–2, 4)

EUROPEAN CONGRESS ON THE INFLUENCE OF AIR POLLUTION ON PLANTS AND ANIMALS, FIRST (1969). *Air Pollution*. 415 pp. Centre for Agricultural Publishing and Documentation, Wageningen. (Chs. 2, 5, 7)

FERRY, B. W., BADDELEY, M. S. and HAWKSWORTH, D. L., eds. (1973). *Air Pollution and Lichens*. 389 pp. Athlone Press of the University of London, London. (Chs. 2–6)

GILBERT, O. L. (1970). Further studies on the effect of sulphur dioxide on lichens and bryophytes. *New Phytol.*, **69**, 605–27. (Chs. 2, 5)

HALE, M. E. (1974). *The Biology of Lichens*. Second edition. 181 pp. Arnold, London. (Ch. 1)

HAWKSWORTH, D. L. (1971). Lichens as litmus for air pollution: a historical review. *Internat. J. Environ. Stud.*, **1**, 281–296. (Chs. 2, 5)

HAWKSWORTH, D. L., ed. (1974). *The Changing Flora and Fauna of Britain*. (Systematics Association Special Volume No. 6.) 461 pp. Academic Press, London and New York. (Chs. 3–7)

HAWKSWORTH, D. L. (1974–6). Literature on air pollution and lichens I–IV. *Lichenologist*, **6**, 122–125; **7**, 62–66, 173–177; **8**, 87–91. (Chs. 2–6; now appearing twice yearly)

HAWKSWORTH, D. L. and ROSE, F. (1970). Qualitative scale for estimating sulphur dioxide air pollution in England and Wales using epiphytic lichens. *Nature, Lond.*, **227**, 145–148. (Ch. 5–6)

LE BLANC, F. and DE SLOOVER, J. (1970). Relation between industrialization and the distribution and growth of epiphytic lichens and mosses in Montreal. *Can. J. Bot.*, **48**, 1485–1496. (Ch. 5)

O'HARE, G. P. (1973). Lichen techniques of pollution assessment. *Area*, **5**, 223–9. (Ch. 5)

PUCKETT, K., NIEBOER, E., FLORA, W. and RICHARDSON, D. H. S. (1973). Sulphur dioxide: its effect on photosynthetic ^{14}C fixation in lichens and suggested mechanisms of phytotoxicity. *New Phytol.*, **72**, 141–54. (Ch. 2)

SMITH, D. C. (1973). *The Lichen Symbiosis*. (Oxford Biology Readers No. 42.) 16 pp. Oxford University Press, London. (Ch. 1)

SWEDEN'S CASE STUDY FOR THE UNITED NATIONS CONFERENCE ON THE HUMAN ENVIRONMENT (1971). *Air Pollution across National Boundaries. The Impact on the Environment of Sulfur in Air and Precipitation*. 96 pp. Royal Ministry for Foreign Affairs and Royal Ministry of Agriculture, Stockholm. (Ch. 7)